New Manager AI Handbook:

The Definitive Detailed Guide to Leading Teams and Leveraging AI for Maximum Productivity

Jazper Carter

i

Cybersoft Publishing LLC

Fort Washington, MD 20744

Drclaude.net

First Edition March 2026

Table of Contents

Foreword

Something unprecedented is happening to managers right now. It is not simply the arrival of artificial intelligence as a productivity tool that would be manageable on its own. What is genuinely new is the collision of two demanding transitions happening simultaneously: the already-difficult leap from individual contributor to people leader, and the obligation to navigate an AI-powered workplace that rewrites workflows faster than most organizations can update their policies.

A first-time manager in 2026 must learn how to run a one-on-one that actually builds trust, conduct a performance review that motivates rather than demoralizes, delegate with enough clarity that work comes back right the first time—and, in the same week, evaluate which AI writing assistant belongs in the team's stack, help a skeptical direct report understand that AI is not coming for their job, and make sense of a data dashboard that did not exist six months ago. That is an extraordinary ask. The truth is that most organizations are not set up to help their newest managers carry it.

This book exists because that gap needed to be closed.

The New Manager AI Handbook was written for three kinds of readers. The first is the newly promoted manager who is weeks or months into the role and already suspects that the mental model they used to succeed as an individual contributor does not transfer cleanly to leadership. The second is the aspiring leader who wants to arrive prepared—someone who understands that the best time to build a management philosophy is before the calendar fills with direct reports. The third is the experienced manager who has led teams for years but senses that the AI era demands a genuine update to their approach, not just a thin technology layer painted over old habits.

If you see yourself in any of those descriptions, this book was written with you in mind.

The dual challenge at the heart of this book—leading people and leveraging AI—is not a contradiction. It is, in fact, a reinforcing loop. Managers who develop strong human-centered skills find that AI amplifies those skills. They can prepare for coaching conversations with a richer context. They can identify performance patterns earlier. They can free up the hours that used to go to administrative overhead and reinvest them in the high-touch work that actually develops people. Conversely, managers who try to use AI tools without the underlying leadership foundation tend to automate mediocrity: faster communication that still misses the point, dashboards that track the wrong things, and decisions that move quickly but land badly.

The chapters ahead are organized around that reinforcing loop. You will build the foundation first: your identity as a leader, your communication style, and your decision-making framework. You will then develop the core management competencies: delegation, performance management, hiring, coaching, conflict management, and personal productivity. From there, you will learn how to apply AI tools with precision to each of those competency areas. And finally, you will look forward to the strategic horizon—culture, transformation, and continuous growth.

This is not a book that asks you to become a technologist. It asks you to become a better leader while also giving you access to powerful tools. The distinction matters. The managers who thrive in the decade ahead will not be the ones who adopt the most tools—they will be the ones who stay relentlessly focused on people, clarity, and outcomes, and use technology only in service of those priorities.

You are not alone in this. Every effective manager working today is figuring out the same terrain. The path is navigable. The skills are learnable. And the manager you are becoming—one who can develop people and deploy AI with equal confidence—is exactly what your team needs.

Let's get started!

1 Introduction

Before you read a single chapter of this book, it is worth spending a few minutes understanding how it is structured, what it assumes about you, and how to get the most out of it. This introduction is not a formality; it is a map.

1.1 A Note on Key Terms

Four terms appear throughout this book and warrant precise definition at the outset.

Manager refers to anyone with direct reports and formal accountability for a team's performance. This includes first-time managers, team leaders building management muscles, and mid-level managers overseeing other managers. The core principles apply at every level, though the context shifts as the scope expands.

Leadership is treated here as distinct from management, though inseparable from it. Management is the set of processes—goal-setting, delegation, performance reviews, and resource allocation—that keep a team functioning. Leadership is the set of behaviors, vision, trust, psychological safety, and inspiration that make a team worth being part of. Great managers develop both. This book addresses both in that order because the processes provide scaffolding on which the behaviors can operate consistently.

AI augmentation describes the use of artificial intelligence tools to extend a manager's capabilities without replacing human judgment. The word augmentation is deliberate. AI, in this context, is not a substitute for knowing your team, making hard calls, or having difficult conversations. It is a force multiplier for managers who already know what good looks like. Throughout this book, AI augmentation is positioned as a practical productivity and decision-support capability—not as a transformation of the management role itself.

Productivity, as used here, means the ratio of meaningful output to time and energy invested—at the individual, team, and organizational level. This definition matters because it is easy to mistake busyness for productivity. A manager who attends 12 meetings and answers 200 messages in a week is busy. A manager who ensures their team delivers on the right priorities, grows in capability, and maintains the energy to sustain performance over time is productive. This book is relentlessly focused on the latter.

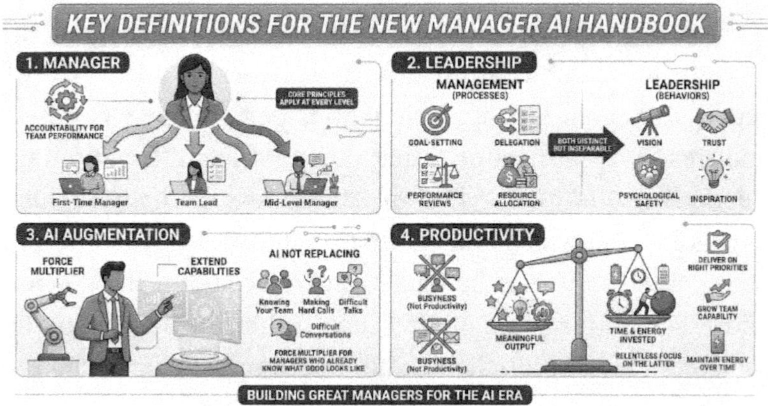

1.2 The Book's Five-Part Structure

The New Manager AI Handbook is organized into five parts, each building on the one before it.

Part One: Foundation (Chapters 1–2) addresses the identity shift from individual contributor to leader. Chapter 1 examines why the transition is harder than it appears and what it demands psychologically and practically. Chapter 2 builds the philosophical and ethical foundation of your leadership approach, your values, your decision-making principles, and your operating model as a manager.

Part Two: Core Management (Chapters 3–5) covers the fundamental processes that separate effective managers from ineffective ones. Chapter 3 is about communication—the skill that underline everything else. Chapter 4 teaches delegation as a discipline, not just a time-saving tactic. Chapter 5 addresses goal-setting and performance management as a continuous system rather than an annual event.

Part Three: People Mastery (Chapters 6–8) goes deeper into the human dimension of the role. Chapter 6 covers hiring and onboarding—the decisions with the highest long-term leverage. Chapter 7 is about coaching, feedback, and talent development—the work that compounds over time. Chapter 8 addresses conflict and difficult conversations—the work most managers avoid and most teams desperately need.

Part Four: AI Productivity (Chapters 9–12) is where technology enters with intention. Chapter 9 establishes personal productivity and time management as the prerequisite for AI leverage. Chapter 10 builds the AI literacy foundation every manager needs. Chapter 11 shows how to deploy AI tools to amplify team output across the management competencies covered in earlier chapters.

Chapter 12 introduces data-driven decision-making as a management practice.

Part Five: Strategic Leadership (Chapters 13–15) looks outward and forward. Chapter 13 covers leading through change and digital transformation—the context in which most managers are now operating. Chapter 14 addresses team culture and engagement as strategic assets. Chapter 15 closes with a framework for continuous growth and future readiness as a leader in the AI era.

1.3 How to Use This Book

This book is designed to work two ways. If you are new to management or new to AI, read it cover to cover. The chapters are sequenced deliberately, each one builds context for the next, and the skills in later chapters assume the foundations laid in earlier ones. If you follow this path, you will finish with a coherent, integrated management philosophy rather than a collection of disconnected tactics.

If you are an experienced manager modernizing your approach, or if a specific challenge is urgent right now, jump to the relevant chapter. Each chapter is written to stand alone, with enough context to be useful without requiring the preceding ones. Use the five-part structure as a navigation guide: identify which domain your current challenge falls under, and go there directly.

Throughout the book, you will find practical frameworks, diagnostic questions, and AI application notes. The frameworks are tools, not rules—adapt them to your context. The diagnostic questions are meant to be answered honestly, not optimistically. The AI application notes connect each management concept to specific ways AI tools can extend your capability in that area.

One final orientation: this book is manager-centric. It is written from the perspective of the person with formal accountability for a team's performance, not from the perspective of a C-suite executive, an HR department, or an organizational theorist. The advice is practical, the examples are grounded, and the standard is not perfection—it is consistent progress toward becoming the kind of manager your team deserves.

2 The Identity Shift: From Individual Contributor to AI-Era Leader

2.1 When the Promotion Changes Everything

It is Monday morning, eight weeks into your new role. You used to be the person your team turned to when the production environment went dark at midnight, when a client escalated at the worst possible moment, or when the product specification made no technical or logical sense. You had answers. You had skills that were tested, refined, and proven under pressure. That mastery was visible, measurable, and deeply satisfying in a way that most people who have never done highly skilled technical or professional work fail to appreciate. The feedback loop was tight and reliable: you built something, shipped something, solved something, and you could see the result. Your identity was grounded in that capability, reinforced every time someone came to you for help and left with a solution.

Today, your calendar looks nothing like that world. It is a wall of one-on-ones, status meetings, planning reviews, and budget conversations. You have a team of seven smart, capable people who are waiting for direction, resources, advocacy with senior leadership, and the kind of operational clarity that allows them to do their best work without having to fight the organization to do it. You have a manager whose expectations of you are almost entirely relational and strategic rather than technical and tactical. You have stakeholders in other departments who need to understand what your team is doing and why it matters to their work. And somewhere beneath all of that, you have a quiet but persistent internal voice that keeps asking: " Am I actually doing anything? Where is the work? Why does nothing feel finished at the end of the day?

This is the moment nearly every new manager describes, regardless of industry, function, organizational size, or technical domain. The scenario is so universal that management researchers have a name for it: the contributor trap. The skills and identity that earned you this promotion are no longer the ones that will make you successful in it. The promotion was a vote of confidence in what you have demonstrated so far. What you need now is something different—a new operating system for how you measure your work, how you invest your time, how you build relationships, and how you define your own success. The good news is that you are not the first to navigate this passage, and the path is well-lit for those willing to follow it.

Diagram 1.1: The Contributor-to-Leader Identity Shift

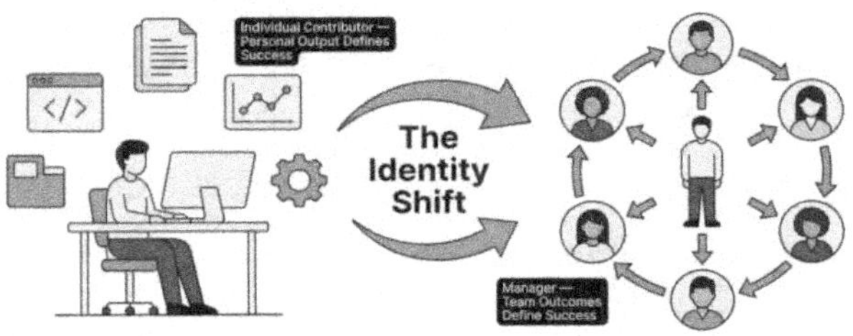

Successfully transitioning from IC to leader requires redefining how success is measured and achieved through the team.

2.2 Why This Shift Is Harder Than It Looks

The contributor to leader transition is one of the most studied inflection points in organizational psychology, and the research is remarkably consistent: most new managers significantly underestimate the disorienting nature of the experience. The disorientation is not a sign of incompetence. It is an accurate, proportionate response to a genuinely novel situation that requires you to operate by different rules, measure yourself against different standards, and build a different kind of daily discipline than the one that made you successful before. Understanding why the shift is hard does not make it easier in the moment—but it does make it possible to respond to the difficulty with curiosity rather than self-criticism, which is a meaningful difference.

The psychological dimension of the transition is real and worth examining directly. Many new managers experience what organizational psychologists describe as role identity conflicts, sustained internal tension between the professional self-concept they have built over years as an expert contributor and the emerging identity of a manager whose value is measured in people and outcomes rather than tasks and deliverables. This conflict manifests in predictable behavioral patterns. The manager steps back in to solve the technical problem because it is faster. The manager who insists on reviewing every piece of work before it goes out because they cannot yet trust the team to share their standards. The manager who stays late doing individual work because it is the only thing at the end of the day that feels concrete and finished. These behaviors are not character flaws. They are the gravitational pull of a deeply ingrained identity that has not yet been replaced by an equally compelling one.

The organizational risk of staying stuck in contributor behaviors is substantial and often invisible to the manager experiencing it. When managers hold individual contribution as their primary mode—doing rather than delegating, solving rather than coaching, optimizing their own output rather than the team's collective capacity—the consequences ripple outward and compound over time. Team members are underutilized, left to execute narrowly defined tasks without the growth opportunities or expanded scope that would develop their skills and engagement. Decisions are bottlenecked at the manager rather than being distributed to the people closest to the relevant information. Strategic work—the planning, relationship-building, and organizational navigation that is, properly, the manager's contribution—gets crowded out by tactical execution that someone else on the team could handle. And the manager, paradoxically, becomes less visible to senior leadership precisely because they are buried in work that should have been delegated, while the genuinely managerial contributions that earn organizational standing are going undone.

For organizations that are actively investing in AI adoption and digital transformation, the stakes are even higher. Managers who cannot operate comfortably at the strategic and people-focused level required by the role are not positioned to guide their teams through the decisions, adaptations, and cultural shifts that AI integration demands. Determining which workflows to automate and which require sustained human judgment, helping team members whose roles are evolving to develop new skills and find renewed purpose, maintaining the ethical guardrails that prevent AI tools from introducing bias or risk into consequential processes—all of these require a manager who has genuinely made the identity shift. The failure to make that shift is not just a personal career limitation. It is a mission-aligned risk for the organization.

Managers staying in individual contributor mode hinder team development and expose the organization to strategic risk.

2.3 Understanding Your New Role

The manager's role can be understood through three core functions that are distinct from, though deeply informed by, individual contribution. The first function is direction-setting: ensuring that your team has a clear, shared understanding of where they are going, why it matters, how their work connects to the organization's mission, and what success looks like in concrete, observable terms. Direction-setting is not just announcing goals. It translates ambiguous organizational strategy into specific, actionable team priorities. It is maintaining alignment as circumstances change and priorities shift. It is about ensuring that every person on your team, on any given day, can articulate what they are working on, why it matters, and what a good outcome looks like. Teams that lack this clarity waste enormous energy on the wrong things, or on the right things in the wrong sequence, and do so with an exhausting uncertainty that erodes engagement over time.

The second core function is resource allocation: ensuring that your team has the tools, information, access, budget, time, and organizational support they need to do their best work. Resource allocation is often underestimated as a management function because it sounds administrative. In practice, it is one of the highest-leverage things a manager does. Removing a bureaucratic obstacle that has been slowing a team member down for three weeks takes 30 minutes of the manager's time and may recoup 30 hours of the team member's time over the next quarter. Securing access to a critical system, introducing a team member to the right internal expert, or advocating for budget to purchase a tool the team needs—these are not peripheral management tasks. They are the daily work of keeping a team operating at the level of effectiveness the organization expects.

The third core function is people development: actively building the capabilities, confidence, and career trajectory of the individuals on your team so that the team's collective capacity increases over time. People development is the function that new managers most consistently underinvest in, partly because it is the most invisible in the short term and the most powerful in the medium term. A team member who grows significantly in their role over eighteen months— who develops skills that expand the team's capabilities, builds relationships that extend the team's organizational reach, and takes on increasing responsibility that allows the manager to focus on higher-order work—represents a form of organizational compounding that no amount of individual output can match. Investing in people development is the clearest expression of the manager's value equation: you are not here to produce the best work yourself. You are here to build the environment and relationships in which the best collective work can happen.

Diagram 1.3: The Manager's Three Core Functions

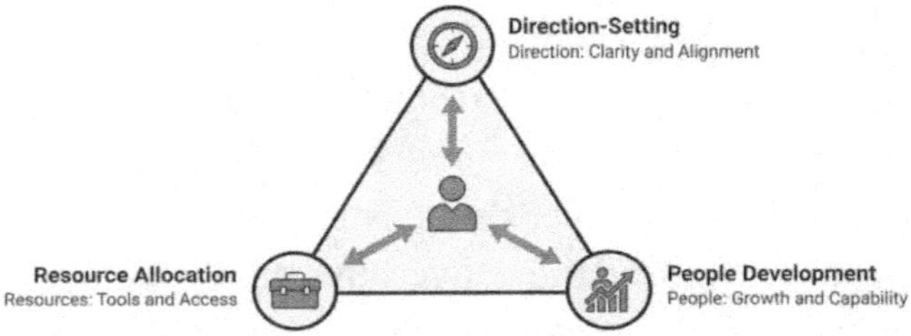

The manager's three chamy a ctront to share gds, manager and slatatenk, and amd development.

2.3.1 The Manager's Value Equation in the Age of AI

AI tools have already changed the economics of individual contribution across most knowledge-work fields, and the pace of that change is accelerating. Tasks that once required hours of skilled human effort—drafting professional communications, synthesizing large volumes of data, generating code scaffolding, summarizing meeting notes, researching competitive intelligence, creating first-pass presentations—can now be completed in minutes with the right tools used thoughtfully. This transformation raises a question that every manager in an AI-augmented organization should think about clearly and honestly: if AI can so dramatically accelerate individual output, what distinctive value does a human manager bring to a team?

The answer is found precisely in the capabilities that AI cannot replicate at the level of quality, judgment, and relational depth that high-performing teams require. Contextual judgment—the ability to weigh competing priorities against organizational culture, interpersonal dynamics, historical context, and genuine mission-alignment—remains a deeply and irreducibly human function. The AI system does not know that two of your team members are in a period of friction that could undermine collaboration on the project it just summarized. It does not know that the organizational priority it is optimizing for is in tension with an informal commitment your team made to another department. It does not know that the team member whose output metrics are strong is quietly burning out and will leave if the work environment does not change. These are not gaps that better data will close. They are the inherent domain of human judgment applied in real relationships over time.

Trust-building—the foundation on which all high-performance team behavior rests—requires sustained physical or virtual presence, emotional attunement, behavioral consistency, and genuine investment in individuals as people rather than as resources. No AI system can replace the relationship between a manager and a team member navigating a difficult stretch: a performance challenge, a personal crisis, a career uncertainty, or a technical failure that has shaken their confidence. Strategic sense-making—the ability to translate ambiguous signals from senior leadership, customers, market dynamics, and organizational culture into coherent, motivating direction for a team—depends on a kind of wisdom that emerges from accumulated experience, institutional memory, and deeply contextual understanding that current AI systems cannot replicate.

The manager who understands this equation clearly uses AI to compress the administrative and analytical dimensions of their role—scheduling, status reporting, data synthesis, first-draft communications, routine documentation—and deliberately invests the recovered time into the high-value human functions: genuine coaching conversations conducted without distraction, thoughtful strategic planning that benefits from the manager's full cognitive presence, intentional relationship-building across the organization, and the kind of deep listening that surfaces the team issues and individual needs that drive real engagement. This is not a speculative vision of an AI-integrated future workplace. It is a workflow-level restructuring available to managers who choose to pursue it deliberately today.

2.3.2 Common First-Time Manager Pitfalls

The mistakes that derail new managers are consistent enough across organizations, industries, and roles that they can be treated almost like a canonical list. Knowing them in advance does not guarantee you will avoid them—the conditions that produce them are deeply human—but it dramatically improves the odds that you will recognize them when they occur and respond with intention rather than doubling down on the behavior. The patterns that follow are drawn from decades of management research and the direct experience of thousands of managers who have made this transition before you.

The first and most common pitfall is what experienced management coaches call the rescue reflex: the impulse to jump in and solve a problem yourself rather than coaching the team member through it. The rescue reflex feels virtuous in the moment. The problem gets solved faster. The team member's frustration is relieved. The manager demonstrates competence and helpfulness. What actually happens at the systemic level is considerably less positive. Every rescue implicitly communicates to the team member that you do not fully trust their ability to work through complexity independently—even if that is not your conscious intention. Every rescue deprives them of an opportunity to develop problem-solving capability, to experience the confidence that comes from working through something difficult, and to build the resilience that complex professional work requires. The team member who is rescued consistently becomes the one who brings every problem to the manager, because they have been trained by experience to expect this pattern. And the manager who consistently rescues has no time for anything else because they are effectively doing the work of both their role and their team members' roles.

The second pitfall is approval-seeking, and it is most acute for managers who were promoted from within a peer group they now lead. The social dynamics of being the manager among former colleagues are genuinely complicated, and the desire to be liked, to avoid the discomfort that comes with setting limits, delivering hard feedback, or making unpopular decisions, is entirely natural. The problem is that approval-seeking management is not management in any meaningful sense. It is the performance of collegial relationships at the expense of the leadership function. Teams do not respect managers who sacrifice honesty for harmony; they find it disorienting and ultimately demoralizing because it signals that the manager is prioritizing their own comfort over the team's actual needs. The feedback that is softened into meaninglessness does not help the team member grow. The decision that is deferred indefinitely to avoid conflict does not go away—it compounds. The standard that is not enforced consistently applies to everyone except the person who pushed back.

The third pitfall is meeting overload, which is almost universal among new managers trying to stay connected to everything their team is doing. The instinct is understandable: you are responsible for outcomes you cannot directly control, and meetings serve as the mechanism for oversight and alignment. In practice, a calendar that is perpetually packed with coordination activity leaves no space for the reflective, strategic, and relational work that constitutes the highest-value management activities. The manager who is in meetings from eight in the morning until six in the evening is not performing at the level the role requires—they are merely appearing productive. At the same time, the substance of management (coaching, strategic thinking, relationship-building, problem anticipation) goes undone. Running fewer, shorter, better-designed meetings is one of the highest-leverage time investments a new manager can make in the first sixty days.

The fourth pitfall is neglecting upward and lateral management—the work of representing your team's interests, communicating your team's contributions, and building the organizational relationships that determine how much support, resources, and visibility your team receives. Most new managers focus intensely downward—on their team—and underinvest in the relationships with their own manager, peers in other functions, and senior stakeholders that determine whether the team's good work is recognized, whether the team has the resources it needs, and whether the team's challenges get addressed before they become crises. Upward and lateral management is not self-promotion. It is operational clarity for the organization. Your manager needs to know what your team is working on, what obstacles they are facing, and where they need organizational support. Your peers need to know how to collaborate effectively with your team. Senior stakeholders need to understand the value your team is creating. Making sure that information flows is part of your job.

Diagram 1.5: Four Common First-Time Manager Pitfalls

Awareness of these common traps helps new managers transition effectively to leadership roles.

2.3.3 Redefining Success — From Personal Output to Team Outcomes

One of the most immediately impactful and practically challenging shifts a new manager can make is in the way they measure their own success on a daily and weekly basis. The individual contributor's operational instinct is to close out each day with a visible, concrete list of completed work: features shipped, analyses completed, reports delivered, problems solved. There is a neurological satisfaction to this kind of discrete, task-based accomplishment that is genuinely motivating, and its absence in the early days of a management role is one of the primary drivers of the identity disorientation described earlier in this chapter. The management workday rarely ends with a neat list of finished items. It ends with conversations that moved something forward, decisions that cleared a path for others, relationships that were strengthened, and problems that were prevented rather than solved—none of which show up in a task list.

A useful, practical reframe is to replace the daily task completion lens with a weekly question framework. At the end of each week, ask yourself: Does my team have the clarity they need to do excellent work without unnecessary uncertainty? Are the individuals on my team growing in their capabilities, or are they stagnating? Have I removed more obstacles from their path than I created? Is the environment I am building one in which people feel psychologically safe enough to take initiative, raise concerns honestly, and invest their genuine best effort? These questions are harder to answer definitively than a task list—the feedback is qualitative, relational, and often delayed—and that ambiguity is itself a fundamental aspect of the management adjustment that must be accepted rather than solved.

Operationally, this reframe has immediate and concrete implications for how you structure your time and evaluate your choices. The manager who spends an hour working through a complex problem with a team member—asking questions that surface the team member's reasoning rather than providing the answer directly—has done valuable, measurable work, even if their personal task list is shorter at the end of the day. The productive one-on-one conversation that surfaces a team member's hidden frustration, reconnects them to the purpose of the work, and restores their engagement is a genuine team outcome, even if it never appears in any project management system. The thirty minutes spent preparing thoughtfully for a senior leadership meeting so that your team's work is accurately represented and appropriately valued is a contribution that will pay forward in resources and recognition for months. Learning to recognize and value these contributions—to feel the satisfaction of management work rather than just the absence of contributor work—is a central developmental task that unfolds over months, not days.

Diagram 1.6: Redefining Daily Success — Contributor vs. Manager Lens

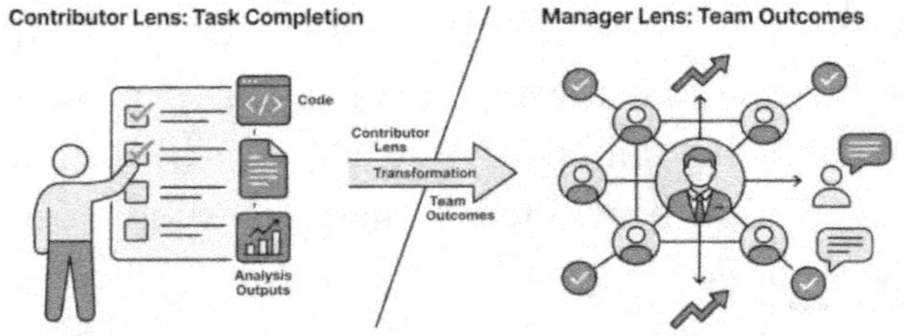

Transitioning to a manager requires a fundamental shift in mindset, from prioritizing personal task completion to empowering the team to achieve collective goals.

2.4 Building Credibility Without Being the Expert in the Room

Among the most anxiety-producing aspects of the transition to management is the realization—often sudden and uncomfortable—that you will regularly find yourself in rooms where other people know more about specific domains than you do. On a high-performing team, this is not just inevitable; it is the intended design. Your role is not to maintain a personal monopoly on expertise. It is to develop, direct, and deploy others' expertise in the service of a shared mission. The manager threatened by team members who know more about a technical domain, a market, a process, or a tool than they do is the one who will, unconsciously, suppress the very expertise that the team's performance depends on. The manager who is genuinely comfortable with their team members being the smartest people in the room on specific topics—and who actively creates conditions for that expertise to surface and be acted on—is building the kind of team that consistently outperforms its apparent capability.

Credibility as a manager is earned through a fundamentally different currency than credibility as an individual contributor, and failing to understand this difference is one of the most common and costly mistakes new managers make. Individual contributor credibility is primarily expertise-based: you are credible because you can do the work better, faster, or more reliably than others. Management credibility is primarily character and behavior-based: you are credible because you are consistent, transparent, respectful, and honest. Consistency means that your commitments, standards, and stated priorities align with your observable behavior day after day—that you do what you say you will do, that you hold standards evenly rather than selectively, that your mood on a given morning does not determine whether you are approachable today. Transparency means that you share the context and reasoning behind decisions rather than just delivering directives—that people on your team understand not just what they need to do but why it matters and how it connects to broader organizational purpose.

Credibility also depends critically on intellectual honesty about the limits of your knowledge. The instinct to project confidence by performing expertise you do not actually have is one of the most reliable credibility-destroying behaviors a manager can engage in. Teams develop finely tuned detection systems for bluffing over time, and the manager who is caught pretending to knowledge or certainty they do not possess loses significantly more credibility than they would have lost by saying clearly, 'I do not know the answer to that—let me find out and come back to you.' Intellectual honesty, particularly from a position of authority, demonstrates the confidence and security that genuine leadership requires. It signals to the team that the manager's stated certainties can actually be trusted, because they are not in the habit of stating certainties they do not have.

In the AI-augmented workplace, this plays out with particular relevance. Many managers are and will increasingly be leading teams that include individuals with deeper technical knowledge of AI tools, systems, and applications than the managers themselves possess. This is not a gap to be anxious about or a problem to be solved by intensive self-study. It is an organizational resource to be designed into the team's decision-making processes. The manager who creates deliberate space for technical expertise to surface and influence how the team adopts and governs AI tools—who asks genuine questions rather than performing understanding, who credits and amplifies the insights of the technically sophisticated team members, and who provides the organizational context and mission-alignment that shapes how that expertise is applied—is performing excellent management. The fact that they could not write the code or configure the model themselves is entirely beside the point.

Diagram 1.7: Credibility Builders vs. Credibility Traps

Credibility Builders	Credibility Traps
Consistency — Do What You Say	Performing Expertise You Lack
Transparency — Share Context and Reasoning	Micromanaging
Intellectual Honesty — Admit Uncertainty	Inconsistent Standards
Respect — Credit and Amplify Others	Ignoring Team Input
Follow-Through — Commitments Kept	Taking Credit Without Giving Credit

Consistent, transparent leadership builds trust, while micromanaging and taking unearned credit destroys it.

2.5 How AI Tools Reshape What Managers Do

The most grounded and immediately useful way to understand how AI changes the practice of management is to begin with an honest accounting of where a typical manager's time actually goes. Ask most managers how they spend their week, and they will describe a set of high-value activities: coaching team members, making strategic decisions, building cross-functional relationships, and planning for the team's development. Examine their calendars and time logs, and what emerges is a considerably different picture. The majority of management time in most organizations is consumed by meeting preparation, attendance at meetings with unclear purpose or outcomes, administrative documentation, performance and status reporting, responding to and drafting communications, and the coordination overhead that results from fragmented tools and information systems. A relatively small fraction of the week—in many cases, less than twenty percent—is spent on the activities that directly constitute management value: genuine coaching, strategic planning, relationship investment, and the kind of proactive problem identification and resolution that prevents issues from becoming crises.

AI tools can now handle significant portions of the first set of activities with a level of quality that exceeds what most managers could produce in the same time. Meeting preparation—researching participant backgrounds, synthesizing recent project context and history, generating agenda templates aligned with stated objectives, and identifying relevant prior decisions and commitments—can be supported by AI tools in a fraction of the time. Meeting notes, action item lists, and follow-up summaries can be generated automatically from recordings or transcripts, reviewed and refined by the manager in minutes, and distributed with a professional quality that would previously have required careful manual composition. Status reporting dashboards can be connected directly to project management data sources, updated automatically, and formatted for different audiences without requiring manual data gathering and synthesis. Draft communications—including sensitive ones, such as difficult feedback conversations, change announcements, or stakeholder escalations—that benefit from a thoughtful first pass before human refinement can be initiated, using AI writing assistance that provides structure, tone guidance, and completeness without replacing the manager's contextual judgment and authentic voice.

This is emphatically not a recommendation to remove human judgment from any of these activities. It is a recommendation to restructure how human judgment is applied within them, and to direct that restructuring toward the activities where human judgment is genuinely scarce and uniquely valuable. The manager who reviews an AI-generated meeting summary for accuracy, completeness, and appropriate tone is exercising relevant professional judgment far more efficiently than the manager who produces that summary from scratch. The manager who refines an AI-drafted communication to reflect their authentic voice, their knowledge of the specific relationship dynamics at play, and their contextual understanding of the organizational situation is operating at a genuinely higher level than the manager who spent twenty minutes staring at a blank page before typing the same draft. The quality of human judgment applied to AI output is consistently better than judgment applied to a blank page, because reviewing and refining is cognitively easier than generating from nothing. The cognitive resources freed up by that difference can be reinvested in higher-order thinking.

The workflow-level impact of systematically applying this restructuring, when done with intention and continuity rather than as a one-off experiment, is substantial. Managers who implement AI assistance for their administrative and analytical tasks and who track the impact honestly report recovering between 5 and 12 hours per week previously consumed by low-leverage administrative tasks. Even at the conservative end of that range, five hours per week represents a 25% expansion in the effective management capacity of a 40-hour workweek. Directed into genuine coaching conversations, strategic planning with senior leadership, relationship-building across the organization, and the proactive team development work that typically gets crowded out by urgency, five additional high-quality hours per week represents a meaningful transformation in what kind of manager you can be—and in the career visibility and organizational impact that management roles require to advance.

Diagram 1.8: AI-Assisted Time Reallocation for Managers

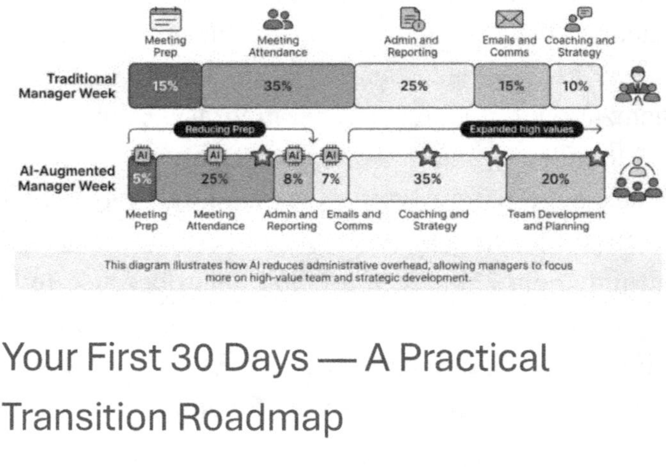

This diagram illustrates how AI reduces administrative overhead, allowing managers to focus more on high-value team and strategic development.

2.6 Your First 30 Days — A Practical Transition Roadmap

The first thirty days in a management role are among the most consequential in a management career. Yet they are consistently underutilized by new managers who lack a deliberate framework for their use. The most common pattern is reactive: attending the inherited meeting cadence, handling what arrives in the inbox, working to avoid visible mistakes while absorbing the institutional and interpersonal landscape. This approach is understandable—the volume of new information in a new management role is genuinely overwhelming—but it carries a real cost. The relational and credibility foundations that will determine your effectiveness for the next year or more are being laid in these thirty days, whether you intend it or not. Making that foundation laying deliberate rather than accidental is one of the highest-leverage investments you can make.

The first ten days should be devoted almost entirely to listening. Schedule one-on-one conversations with every direct report, your own manager, key peers whose teams interact with yours, and the major internal stakeholders your team serves. In each conversation, ask substantially more than you tell. The questions that matter most are open-ended and genuinely curious: What is working particularly well right now, and why? What is most frustrating about the current way of working? What does this team contribute that the organization would most miss if it were to disappear? What is the biggest obstacle between where the team is today and where it could be? What would you most want a new manager to understand about this role, this team, or this organization that might not be obvious from the outside? What does the team need from a manager that it may not have been getting? Take detailed notes, identify recurring themes, and, with discipline, resist the impulse to offer solutions, process improvements, or assessments immediately. The listening phase has no value if the performing phase contaminates it.

The second ten days should begin the process of synthesis, orientation, and structured learning. Review any available documentation: team charters, project plans, performance data, decision logs, prior retrospectives, customer feedback, and organizational strategy documents. Identify the three to five decisions that are most urgent and most properly yours to make with the information currently available. Identify the assumptions your team is operating on that seem most in need of explicit testing. Confirm your synthesis of the team's current priorities with your own manager, checking that your read of what matters most aligns with organizational expectations. Begin deliberately establishing the communication cadences that will define your management rhythm: the structure of your team meetings, the frequency and format of your one-on-ones, the channels and norms for async communication, and the mechanisms for upward and lateral reporting. These are not bureaucratic overhead. They are the operational infrastructure of management effectiveness.

The final ten days of the first month should produce three concrete, tangible outputs that signal clearly to your team and the organization that you are operating as a manager, not as a visitor learning the landscape. The first output is a written document, even a brief one, that articulates your current understanding of the team's mission, its most important priorities, and the key challenges that will require attention over the next quarter. Share this document explicitly with your team and your manager to solicit feedback, corrections, and alignment. The act of writing it and sharing it forces a level of clarity that internal synthesis alone does not produce, and it gives your team and manager a concrete artifact to respond to rather than a general impression. The second output is at least one clear decision or process adjustment that directly addresses a friction point your team identified during the listening phase. This demonstrates that the listening was genuine and that your presence in the role will translate into tangible improvements in the team's working environment. The third output is an explicit, direct conversation with each direct report about your working style, expectations, and what they can count on from you. This is not a formal document or a performance conversation. It is a human conversation that begins to define the relationship and signals that you are present, engaged, and intentional about how you lead.

Diagram 1.9: The 30-Day New Manager Transition Roadmap

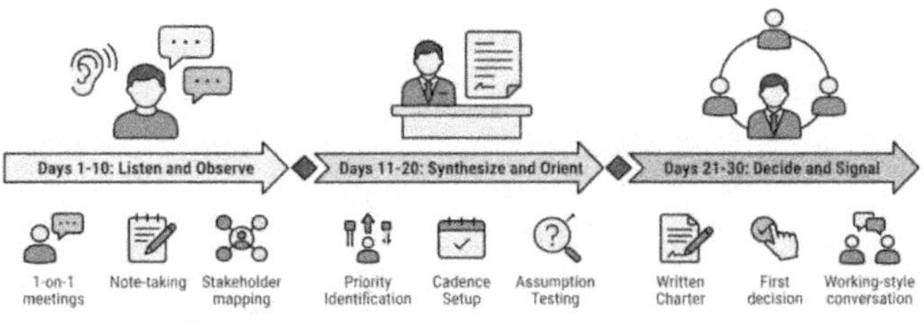

A structured approach to successfully integrate new managers into their role and team.

2.7 Building Your AI Foundation in the First 90 Days

The first 90 days of a management role offer a specific, time-sensitive opportunity to establish your approach to AI tools in a way that will shape how your team thinks about and uses these technologies for years to come. Most new managers either ignore AI integration entirely in the early months—too focused on relationship-building and role stabilization to add another dimension—or adopt tools reactively and inconsistently, without the deliberate framework that produces lasting team behavior. Neither approach serves the manager or the team well. A more effective approach treats AI tool adoption as a core element of establishing your management practice, with the same intentionality you bring to defining your communication cadence or establishing your feedback culture.

Begin by understanding the AI tools your team members are already using, formally or informally. You may find a wide range of adoption: some team members enthusiastically using AI writing, coding, or research tools daily; others skeptical or unaware; and some operating under a misunderstanding of what organizational policy permits. Getting a clear picture of the current state, without judgment or premature standardization, is the necessary foundation for deliberate integration. Team members who have been using AI tools effectively often have insights about which workflows benefit most from assistance and which require sustained human judgment—insights that will be more valuable than any external research you might conduct.

From that baseline, identify two or three specific workflow categories where AI assistance offers clear, measurable time savings or quality improvements for your team. These categories should be ones where the output is reviewable and where errors are detectable—where AI assistance genuinely compresses the time to a good result rather than introducing uncertainty about whether the result is good at all. Common high-value categories for management-level AI adoption include meeting summarization and action item tracking, communication drafting and revision, data synthesis and reporting, and research and competitive intelligence. For each category, establish a simple team protocol: which AI tools are available and approved, how outputs should be reviewed and labeled, and what the quality standard is for AI-assisted outputs. These protocols do not need to be elaborate. A one-page team AI use guide, shared, discussed, and revised as the team learns, is more valuable than a comprehensive policy that no one reads or applies.

Finally, model the behavior you want to see. If you want your team to use AI tools thoughtfully—using them for acceleration and first-pass quality, while maintaining human judgment over final outputs—you need to be visibly doing exactly that. Share the fact that you used an AI writing assistant to draft the project update, which you then revised significantly. Acknowledge that the meeting summary was AI-generated and that you reviewed and corrected it before distributing it. Discuss openly in team forums what worked, what did not, and what you learned. The manager who leads AI adoption by modeling it with transparency and intellectual honesty creates a team culture of responsible experimentation that is genuinely difficult to build any other way. This is what it means to be responsible by design rather than compliant by policy.

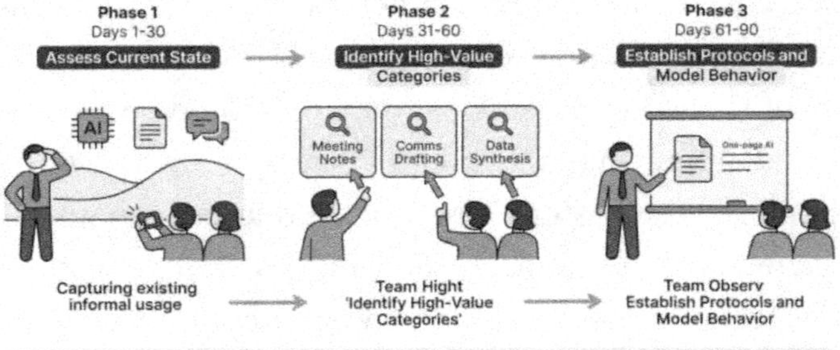

2.8 Manager's Checklist — The Identity Shift

Use this checklist during your first sixty days and return to it quarterly. These are not one-time orientation tasks. They are ongoing practices that compound in effectiveness over time as they become habitual and as your team comes to rely on them.

2.8.1 Mindset and Identity

- Write down the three individual contributor behaviors you are most likely to fall back on under pressure—rescue reflex, approval-seeking, technical solutioning—and create a personal protocol for noticing when you are doing it and redirecting toward the manager behavior it should be replaced with.

- Identify one or two team members for each of your core technical strengths whose expertise you can visibly rely on, publicly credit, and actively create space for in team discussions and with senior stakeholders, rather than positioning yourself as the primary expert.

- Rewrite your personal definition of a successful workday and a successful week in explicitly managerial terms—team clarity, obstacle removal, coaching conversations, strategic planning—rather than personal task completion.

- Schedule a direct conversation with your manager about what the first sixty days should produce and what specific observable behaviors will signal that you are making the identity transition effectively.

- Identify the outcome you are most tempted to control directly rather than delegate, and practice articulating clearly why that outcome is better served by investing in the team member's ownership than in your own execution.

2.8.2 Role and Relationships

- Complete a one-on-one conversation with every direct report within the first ten days, structured primarily around listening and understanding rather than directing, informing, or assessing.

- Map your key stakeholders across three dimensions— upward to your manager and senior leadership, lateral to peer managers and cross-functional partners, and external to key clients or organizational dependencies—and establish a regular communication rhythm with each.

- Identify the three most consequential decisions that are waiting for your input or authority, make them explicit with visible reasoning, and communicate the decisions and rationale to the relevant team members.

- Produce a written team charter or priority document within the first thirty days, share it with your team and manager for reaction and alignment, and revise it based on the input you receive.

- Have an explicit working agreement conversation with each direct report: how you prefer to communicate, what you expect in terms of transparency and escalation, and what they can specifically count on from you as their manager.

2.8.3 AI and Workflow Integration

- Conduct a structured personal time audit for your first two weeks, categorizing every significant time investment, identifying the pattern of how your time is actually distributed, and comparing it to how you would choose to distribute it ideally.

- Survey your team members about the AI tools they are currently using—formally or informally—the workflow categories those tools support, and their assessment of where AI assistance creates the most genuine value.

- Identify two administrative or analytical tasks that currently consume significant management time and research one AI tool or workflow approach that can compress the time required for each by at least fifty percent.

- Pilot one AI productivity application—meeting summarization, scheduling assistance, communication drafting, or status report generation—during a defined two-week window and track the time recovered and the quality of output compared to your prior approach.

- Draft a brief team AI use guide (one to two pages) that articulates which AI tools are available and approved for team use, how AI-assisted outputs should be reviewed and labeled, and what the expected quality and transparency standards are.

2.9 What to Carry Forward

The transition from individual contributor to manager is not a single decision made on the day of the promotion. It is a sustained and iterative identity reconstruction that unfolds over months, expressed in the accumulation of small, daily choices: the decision to coach rather than rescue, to ask rather than tell, to measure your day's success by what your team accomplished rather than what you personally produced. The managers who navigate this transition most successfully are not the ones who feel no disorientation—everyone feels it. They are the ones who develop the discipline to act well toward the new identity even when the old one's gravitational pull is strong, and who build the capacity to find genuine satisfaction in the harder-to-see, slower-to-materialize outcomes that management produces.

AI represents a genuine structural shift in the practice of management, not a set of optional tools that forward-thinking managers might choose to explore. The time it can recover, the quality it can improve, and the administrative complexity it can absorb are not marginal benefits. They are the operational foundation that enables managers to spend most of their professional time on the distinctively human work of creating teams, building careers, and driving organizational performance. The manager who treats AI integration as a core part of their management practice from the beginning—establishing protocols, modeling responsible use, and tracking the workflow-level impact—is not just being efficient. They are building the team culture and organizational capabilities that will compound in value over the years.

Every manager before you has crossed this same threshold, in some form, in some version of this identity challenge. The specific technologies are new; the fundamental transition is not. Some made the crossing smoothly, some stumbled and recovered, and all of them learned something essential: the skills that earned you this opportunity are the foundation, not the ceiling. What you build on that foundation—the relationships, the judgment, the leadership practice that develops only through the accumulated experience of leading people through real challenges—is what will define your management career. You are not the first to feel the disorientation of a job that has changed its rules on you. You are precisely in the right place to begin building what comes next.

3 Building Your Leadership Foundation

3.1 The Moment the Mirror Comes Out

Three months into her first management role, a product manager named Jordan sat in a one-on-one with a direct report who was struggling to meet deadlines and growing visibly disengaged. Jordan came prepared with a performance framework, a set of expectations she wanted to clarify, and a plan to reset the working agreement. She left the conversation thirty minutes later without having said most of it, because the team member had spent most of the meeting describing his experience of working on Jordan's team, and what he described was almost the opposite of what Jordan believed she was doing. He said she communicated priorities inconsistently. He said she changed direction abruptly without explanation. He said that in moments of pressure, she became less approachable rather than more, and that the team had learned to bring problems to her only when they had no other option. Jordan genuinely believed she was a transparent, accessible, and steady manager. The feedback was not an attack. It was a mirror. And what she saw in it required a fundamental revision of her understanding of herself as a leader.

This kind of moment—the jarring collision between the manager you believe yourself to be and the manager your team is actually experiencing—is not a sign of failure. It is a sign that the developmental work of building a genuine leadership foundation has begun in earnest. The foundation is not built solely on frameworks and techniques, important as those are. It is built from self-knowledge: an honest, empirically grounded understanding of your natural tendencies, your default responses under pressure, your blind spots, and the gaps between your intentions and your observable behavior. Without that foundation, every leadership technique you apply is being applied to a construction you have not fully examined. With it, you can build deliberately and with precision.

AI-powered assessment tools have made self-examination more accessible and rigorous than at any point in management history. Personality frameworks that once required trained facilitators to administer and interpret are now available on platforms that deliver results in minutes, contextualized for management roles, and mapped to team effectiveness research. Three-hundred-sixty-degree feedback systems that once required weeks of manual survey distribution and analysis now run continuously, providing managers with regular, quantitative, and qualitative data on how their leadership behavior is landing with the people who experience it. None of these tools replaces the human work of growth and reflection. But they accelerate the feedback loop, significantly compressing the developmental timeline.

Diagram 2.1: The Leadership Foundation Framework

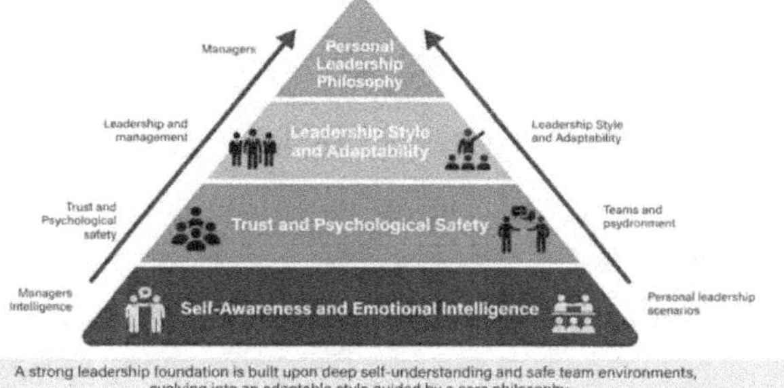

A strong leadership foundation is built upon deep self-understanding and safe team environments, evolving into an adaptable style guided by a core philosophy.

3.2 Why the Leadership Foundation Determines Everything Else

The management literature is extensive and consistent on one empirical finding that surprises most new managers: the technical quality of the systems, frameworks, and processes a manager implements matters far less than the quality of the leadership relationships within which those systems operate. A team led by a manager with high self-awareness, genuine emotional intelligence, and a well-calibrated leadership style will find ways to compensate for imperfect systems, unclear processes, and resource constraints. A team led by a manager who lacks those foundations will underperform even with excellent systems, because the psychological environment those foundations create—or fail to create—is the variable that most strongly determines whether team members invest their full capability, take the risks that innovation requires, and communicate with the honesty that effective collaboration demands.

This is not a soft or peripheral finding. It is operationally central. The research on psychological safety conducted by Amy Edmondson at Harvard Business School, replicated across hundreds of teams in multiple industries and organizational contexts, establishes clearly that teams whose members feel safe to speak up, admit mistakes, ask questions, and challenge assumptions significantly outperform teams that lack this safety on every measurable dimension of performance: innovation, quality, efficiency, and employee retention. Psychological safety is not a cultural amenity. It is a performance mechanism. And it is created or destroyed primarily by the manager's behavior, day by day, and interaction by interaction. The manager's emotional intelligence—their ability to perceive, understand, and regulate their own emotions and to accurately read and respond to others' emotional states—is the primary determinant of whether they build psychological safety or erode it.

The AI dimension adds both new tools and new urgency to this foundation-building work. New tools, because AI-powered assessment platforms can now provide managers with more continuous, more specific, and less biased data about their leadership behavior than any previous methodology has allowed. New urgency: teams navigating the complexity and uncertainty of AI integration require psychological safety in greater quantities than they need for stable, routine work. When team members are asked to learn new tools that may change their roles, experiment with workflows whose outcomes are uncertain, and make judgment calls about AI outputs that carry organizational risk, they need to feel safe asking questions, raising concerns, admitting confusion, and surfacing problems early. The manager who has not built a foundation of trust, emotional intelligence, and adaptive leadership is asking their team to navigate AI transformation on the most difficult possible terrain.

Your leadership foundation is also the primary driver of your own sustainability in the management role. Management is genuinely difficult work. It is relational, ambiguous, emotionally demanding, and rarely generates the tight, visible feedback loops that make individual contribution so satisfying. Managers without a strong foundation in self-awareness and emotional intelligence are vulnerable to the specific stresses of the role, which can manifest as burnout, defensive behavior, escalating micromanagement, and eventual disengagement. The investment in foundation-building is not just about team performance. It is an investment in your capacity to lead effectively and with satisfaction over a management career that, if it goes well, will span decades.

3.3 Leadership Styles and When to Use Them

One of the most durable and practically useful contributions of organizational psychology to management practice is the research on leadership style—the consistent patterns in how managers direct, develop, and motivate the people they lead. The foundational insight, developed and empirically validated by Paul Hersey and Ken Blanchard's situational leadership model and substantially extended by subsequent research, is that there is no single best leadership style. Effective leaders accurately read a team member's developmental level and motivational state in a specific situation and apply the style most appropriate to that combination. The manager who applies the same directive, highly structured approach to an experienced senior professional that they apply to a newly onboarded junior team member is misapplying their tools as surely as someone using a hammer to turn a screw.

The major leadership styles operate along two primary dimensions: the degree of directive behavior (how much the manager specifies what to do, how, and by when) and the degree of supportive behavior (how much the manager provides emotional encouragement, recognition, and relationship investment). In a high-directive, low-support style—appropriate for team members new to a task and needing structure and clarity above all—the manager provides specific instruction, close follow-up, and clear criteria for success. In a high-directive, high-support style—appropriate for team members who have some competence but are lacking confidence—the manager combines clear direction with active encouragement, collaborative problem-solving, and recognition of progress. In a low-directive, high-support style—appropriate for team members who have sufficient competence but may need motivational engagement—the manager asks more than tells, involves the team member in decision-making, and provides emotional investment and recognition while pulling back on task level instruction. In a low-directive, low-support style— appropriate for high-competence, high-confidence team members—the manager delegates fully and primarily monitors rather than directs or supports, trusting the team member's capability and judgment.

Most new managers have a default style that is the product of their own experience of being managed, their personality, and the organizational culture they have been shaped by. Many default to high-directive styles across the board because directive management feels clear and controllable in the ambiguous early days of the role. Others default to low-directive, high-support styles, reflecting a preference for collegial relationships over hierarchical authority. Neither default is wrong in itself; the problem is the default—the application of a single style regardless of context, rather than a conscious calibration to the situation at hand. The developmental goal is style range: the ability to operate authentically and effectively across the full spectrum, reading the situation accurately and deploying the appropriate style without it feeling forced or performative.

Diagram 2.2: Situational Leadership Style Matrix

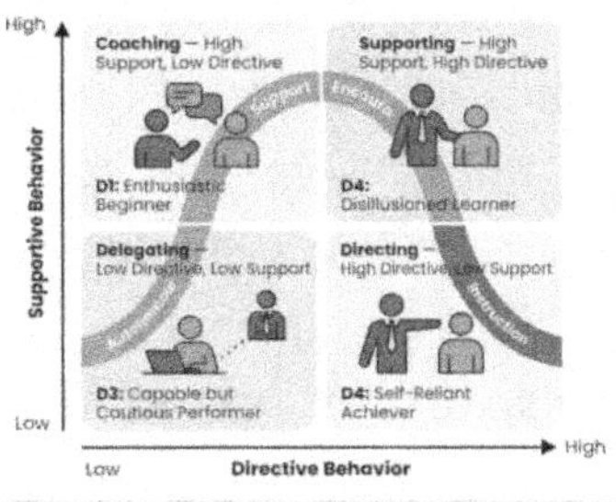

Successful leaders adapt their approach based on the team's competence and commitment levels.

3.3.1 Adapting Style for Diverse Teams and Remote Environments

The complexity of style adaptation increases substantially when teams are geographically distributed, culturally diverse, or operating in hybrid arrangements in which some members are co-located, and others are remote. Physical distance removes the most immediate channels of managerial attunement—the ability to read body language, observe informal interactions, sense the emotional temperature of the team's workspace, and provide the kinds of spontaneous, relationship-sustaining micro-interactions that build psychological safety over time. Cultural diversity adds layers of different norms around authority, directness of feedback, expression of conflict, and appropriate working hours, all of which interact in ways that can make a well-intentioned management style land very differently than intended.

Effective style adaptation for diverse and distributed teams requires two foundational investments. The first is explicit communication of norms and expectations that would remain implicit in a co-located, culturally homogeneous environment. Remote team members cannot observe how decisions are made, how conflict is handled, or what informal behaviors are expected and rewarded—they can only experience the explicit structure the manager creates. The manager who invests in making the team's operating norms explicit, documented, and regularly revisited is removing a significant source of uncertainty and inequity between co-located and remote team members. The second investment is differentiated connection—making a deliberate effort to understand the specific cultural context, working environment, and individual preferences of each team member, rather than applying a culturally specific default and expecting it to translate universally. This is more demanding than uniform management, but it is the only approach that produces genuine equity and high performance across a diverse team.

AI tools offer meaningful support for both investments. Translation and communication assistance tools reduce the friction of cross-language communication and help managers review their written communications for cultural assumptions that may not be universally legible. AI-powered meeting facilitation tools can help ensure participation is distributed among team members with varying comfort levels speaking up in group settings. AI-supported asynchronous communication tools enable remote team members to contribute to decisions and discussions without being disadvantaged by time zone differences. And when applied to team communication data with appropriate privacy protections and team consent, AI sentiment analysis tools can surface signals of engagement, frustration, or disconnection that are much harder for managers to read remotely than in person.

3.4 Emotional Intelligence — The Manager's Superpower

The term emotional intelligence entered popular consciousness through Daniel Goleman's work in the mid-1990s. Its overuse in corporate training materials has unfortunately diluted its meaning to the point that many managers treat it as a personality trait—something you either have or do not—rather than a trainable set of competencies with a direct, measurable impact on management effectiveness. The research is considerably more precise. Emotional intelligence, as defined and measured in organizational research, comprises four distinct competencies that independently and collectively predict management effectiveness: self-awareness, self-regulation, social awareness, and relationship management. Each of these competencies is learnable, improvable, and meaningfully influenced by the feedback and reflective practice that effective managers systematically invest in.

Self-awareness is the foundational competency: the accurate, real-time understanding of your own emotional state, how that state influences your thinking and behavior, and how your behavior is likely to affect the people around you. Many managers believe they are self-aware, meaning they have general insight into their personality and tendencies. Genuine self-awareness in a management context is considerably more demanding: it means noticing, in the moment, that you are feeling defensive in response to a team member's feedback and choosing how to respond to that defensiveness rather than acting from it automatically. It means recognizing that your irritability during a late afternoon meeting is due to fatigue, not to your team member's actual performance. It means understanding that your default mode of communicating confidence can read as dismissiveness to team members who interpret directness as authority rather than openness.

Self-regulation—the ability to manage your emotional states in ways that preserve the quality of your judgment and the safety of your interpersonal environment—is the competency that separates managers who are safe to work for from those who are not. The manager who cannot regulate their frustration when things go wrong creates a team environment where people are cautious, avoid sharing bad news, and invest cognitive energy in managing the manager's mood rather than solving problems. The manager who can absorb pressure, uncertainty, and disappointment while maintaining a steady, approachable presence—without suppressing or performing, but genuinely regulating—creates the psychological safety that allows problems to surface and be solved early. Self-regulation is not emotional blankness. It is emotional intelligence applied to the timing and manner of emotional expression so that it serves the team's functioning rather than disrupting it.

Social awareness—the ability to accurately read the emotional states, needs, and dynamics of others and of the team as a whole—is the competency that transforms the quality of information a manager works with. Managers with high social awareness detect engagement problems, interpersonal friction, and shifts in motivation weeks before they become visible in performance data. They read the room in a meeting—noticing who is disengaged, who has something to say but is holding back, and what the team's collective energy level suggests about the feasibility of the proposal—and adjust their approach accordingly. They understand the informal social structures that shape how information flows and how influence operates within their team. They factor that understanding into how they communicate, assign work, and make decisions.

Relationship management—the ability to use your emotional intelligence to build and sustain the quality of relationships that high-performance work requires—is the competency that manifests most visibly in management outcomes. It includes the ability to inspire and motivate without manipulation, to give feedback that develops rather than diminishes, to navigate conflict in ways that strengthen rather than fracture working relationships, to build the kind of collaborative environment where the whole consistently exceeds the sum of its parts. Relationship management is not naturally soft or interpersonally passive. It includes the ability to have difficult conversations with genuine care—to deliver hard truths in ways that the recipient can hear and act on, to set limits and enforce standards while maintaining a relationship of mutual respect.

Diagram 2.3: The Four Emotional Intelligence Competencies for Managers

These four competencies enable managers to build stronger teams and drive organizational performance.

3.4.1 Using AI-Powered Tools to Build Emotional Intelligence

The application of AI to emotional intelligence development represents one of the most practically significant intersections of technology and leadership development available to managers today. AI-powered platforms can now deliver personality and behavioral assessments with the depth and specificity previously available only through highly trained professional coaches, provide multi-rater feedback collection and synthesis at a frequency that makes development feedback genuinely continuous rather than annual, and analyze patterns in a manager's communication data—with appropriate consent and privacy protections—to identify behavioral tendencies that may not be visible to the manager themselves.

Personality and behavioral assessment platforms such as those built on the Hogan Leadership Forecast Series, the Predictive Index, or the Human Synergistics Circumplex have been integrated with AI-powered development coaching systems that translate assessment results into specific behavioral development targets and provide ongoing nudges, reflective prompts, and progress tracking. For a new manager who may not have access to a formal executive coach or a structured management development program, these platforms provide a meaningful alternative that is both more accessible and more continuous than traditional approaches. The key to using them effectively is to engage with the results actively and skeptically—treating the output as hypotheses about your behavioral tendencies that deserve examination and testing through direct observation, not as definitive descriptions of fixed traits.

Three-hundred-sixty-degree feedback tools—platforms that collect structured feedback from a manager's direct reports, peers, and manager on specific behavioral competencies—have been substantially improved by AI integration. Modern platforms can synthesize open-ended qualitative feedback, identify themes that appear across multiple respondents, flag discrepancies between the manager's self-assessment and others' perceptions, and generate development recommendations that are specific and actionable rather than generic. When implemented with genuine psychological safety—when team members believe their responses are truly anonymous and that the purpose is development rather than evaluation—these tools provide management data that no amount of self-reflection can produce on its own. The perspective of the people who experience your management behavior daily, when captured honestly and analyzed rigorously, is among the most valuable inputs for leadership development.

Diagram 2.4: AI-Powered Leadership Development Tools

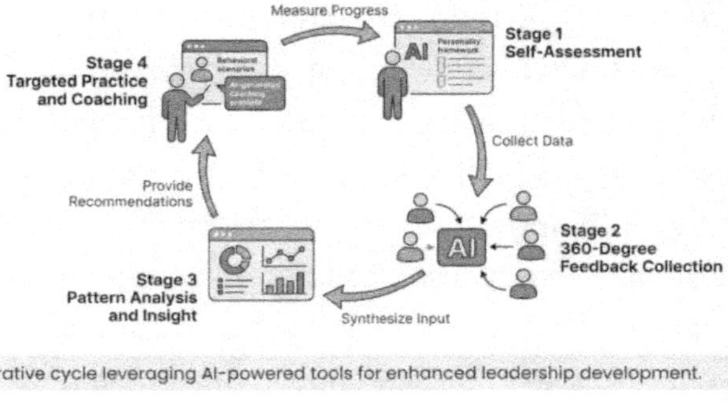

Iterative cycle leveraging AI-powered tools for enhanced leadership development.

3.5 Self-Awareness and Identifying Your Blind Spots

Blind spots—the dimensions of your behavior that are visible to others but not to you—are universal. No manager, regardless of experience level or self-reflection capacity, operates without them. The question is not whether you have blind spots but whether you have the mechanisms in place to discover and address them before they cause the kind of leadership damage that is difficult to repair. The most reliable mechanism is a systematic commitment to feedback: seeking it actively, creating conditions in which people feel safe to provide it honestly, and responding to it in ways that reinforce their willingness to provide it again.

The process of identifying blind spots begins with an honest examination of the gap between your self-assessment and others' assessments of the same behaviors. If you believe you are a strong communicator and your team's feedback suggests that they consistently feel unclear about priorities, the gap between those two data points is a productive area of inquiry. The instinctive response to such a gap is defensiveness—to explain the discrepancy by questioning the accuracy or motivation of others' perceptions rather than examining your own behavior. The developmental response is to get curious: what, specifically, is the team member experiencing when they say their priorities feel unclear? What am I doing or not doing that produces that experience? What would I need to change about my behavior—not about how I think, but about what I visibly do—for the experience to be different?

Common blind spots for new managers cluster around predictable patterns. Managers who were highly effective individual contributors often have a blind spot regarding the pace and complexity of the information they deliver: what feels like necessary context to the manager can feel like overwhelming detail to team members who need clear priorities and simple direction. Managers with high standards and strong quality instincts often have a blind spot in how they communicate those standards: what feels like an appropriate quality expectation to the manager feels like moving goalposts to team members who thought they understood what good looked like. Managers who are naturally decisive and confident often have a blind spot in the space they create for others to contribute. What feels like helpful clarity to the manager can feel like insufficient consultation to team members who want to be involved in decisions that affect their work.

AI-powered feedback platforms offer a capability particularly useful for blind spot identification: they can analyze patterns in large volumes of qualitative feedback data and surface recurring themes across multiple respondents, time periods, and contexts. A single piece of critical feedback is easy to attribute to a difficult personality or a bad day. A pattern that appears in the feedback of three different team members over six months, identified and surfaced by an AI analysis system, is considerably harder to dismiss and considerably more useful as a development target. The most effective managers treat their AI-powered feedback data not as a report card but as a map—an imperfect yet genuinely informative guide to the terrain of their leadership behavior as others experience it.

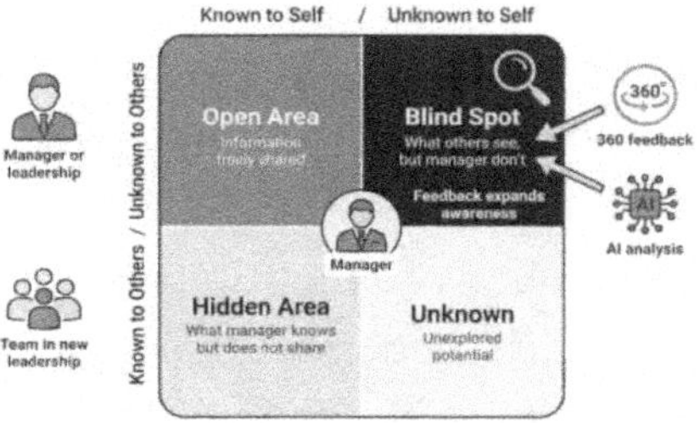

Diagram 2.5: The Blind Spot Discovery Process

Utilizing comprehensive feedback loops and data analysis helps leaders identify unconscious biases and expand their self-awareness.

3.6 Building Trust and Psychological Safety

Trust is the mechanism through which leadership authority is converted into leadership effectiveness. A manager may have formal authority—the organizational position that creates the right to direct work, make decisions, and evaluate performance. Still, if the team does not trust the manager, that authority produces compliance at best and disengagement or active resistance at worst. Trust is not granted by title. It is earned by consistent behavior over time, and it is significantly easier to destroy than to build. Understanding what builds trust and what erodes it—and choosing your behavior accordingly, especially in difficult moments—is among the most important leadership competencies a manager develops.

Research on trust in organizational relationships consistently identifies four primary components: competence trust (the belief that the manager is capable and knowledgeable enough to lead effectively), integrity trust (the belief that the manager is honest and that their stated values align with their observable behavior), benevolence trust (the belief that the manager genuinely cares about the team members' interests, not just the organization's interests or their own career), and consistency trust (the belief that the manager's behavior is predictable—that they will behave in the future as they have behaved in the past). New managers tend to focus most of their trust-building effort on competence, which is understandable given the credential anxiety that accompanies the transition. But research suggests that integrity, benevolence, and consistency have greater predictive power for team performance and retention than competence alone. Teams will work hard for a manager they trust as a person, even if that manager does not have all the technical answers. They will not invest discretionary effort for a manager they believe is dishonest, self-serving, or unpredictable, regardless of technical expertise.

Psychological safety—the shared team belief that it is safe to take interpersonal risks: to speak up with concerns, admit mistakes, ask questions, or challenge the manager's thinking without fear of punishment or humiliation—is built on the foundation of trust and amplified by specific managerial behaviors that signal safety consistently and visibly. The most powerful of these behaviors are: responding to mistakes and failures with curiosity rather than blame (what can we learn from this, rather than whose fault was this); actively inviting dissent and contrary perspectives in discussions and decisions (what is the strongest argument against this approach?); visibly acknowledging and acting on team members' concerns and ideas rather than filtering them through a managerial gatekeeping lens; and modeling vulnerability and intellectual honesty by openly acknowledging your own uncertainties, mistakes, and developmental edges.

The relationship between psychological safety and AI integration is particularly important and often overlooked in AI adoption planning. When organizations introduce AI tools, team members have genuine concerns: about how their roles will change, about the reliability of AI outputs, about what happens if they make an error using an AI-assisted workflow, and about whether the organization intends to reduce headcount. These concerns are not irrational; they deserve transparent, honest engagement rather than reassurance that ignores their legitimacy. The manager who creates a high-psychological-safety environment before and during AI adoption gets teams that experiment with AI tools more genuinely, surface problems with AI outputs more reliably, and develop genuine capability with AI workflows more rapidly. The manager whose team lacks psychological safety gets adoption performance—team members going through the motions of using AI tools while privately distrusting the outputs and hedging every result, which produces no productivity gains while carrying all the risk.

Diagram 2.6: Building Psychological Safety — Manager Behaviors

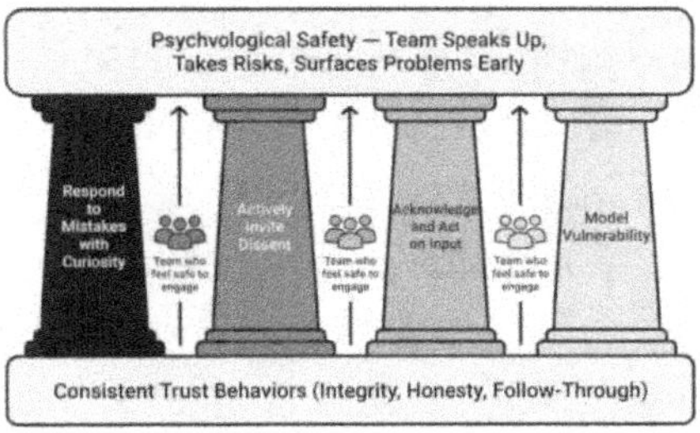

Psychological safety — team speaks engage, that a team speaks Up, takes risks, surfaces problems early.

3.7 Developing Your Personal Leadership Philosophy

A personal leadership philosophy is a clear, written articulation of what you believe about leadership—what it is for, what it demands of the leader, how it should operate in relationship with the people being led, and what values and principles will govern your behavior when the environment is difficult, and the easy path is not the right one. The philosophy is not a mission statement or a list of aspirational traits. It is a working document, regularly tested against experience and regularly revised as your understanding of leadership deepens through practice. Its primary value is not as a communication artifact—something to share with your team or include in a presentation. Its value lies in serving as an anchor for your own behavior in situations where the pressure to act against your values is real, and the temptation to rationalize is strong.

Developing a genuine leadership philosophy requires engaging honestly with several foundational questions that most managers have not explicitly answered. What do you believe leadership is fundamentally for—is it primarily about achieving organizational outcomes, or about developing the people who achieve those outcomes, or about some specific integration of those purposes? What do you believe about the relationship between authority and service—does positional power obligate the person who holds it to serve the people subject to it, or does it primarily confer the right to direct and decide? What do you believe constitutes unacceptable behavior in a leadership role—what lines would you not cross regardless of organizational pressure, personal consequences, or the sophistication of the rationalization on offer? What kind of team culture do you believe produces the best work, and what specific leadership behaviors does building that culture require from you?

The most effective leadership philosophies are developed through an iterative process: begin with draft answers to these foundational questions, test those answers against your experience over two or three months of management practice, and revise them based on what the experience reveals about what you actually believe rather than what you think you should believe. The manager who writes a leadership philosophy on their second day in the role and never revises it has produced a document that reflects their aspirational self-concept rather than their genuine understanding. The manager who writes a draft, leads a team through a genuinely difficult stretch, and then revisits the philosophy in light of that experience has produced something considerably more grounded and considerably more useful.

Share your leadership philosophy with your team—not as a declaration of finished wisdom, but as an invitation to understanding and accountability. When your team knows explicitly what you believe about leadership, what principles you have committed to, and what behaviors you have declared as non-negotiable for yourself, they have a basis for holding you accountable to those declarations. That accountability is not a vulnerability. It is one of the mechanisms through which the consistency that builds trust is created and sustained. The team that has heard their manager explicitly commit to responding to mistakes with curiosity rather than blame will notice—and note with care—when the manager does or does not follow through on that commitment in a difficult moment.

Diagram 2.7: Developing Your Personal Leadership Philosophy

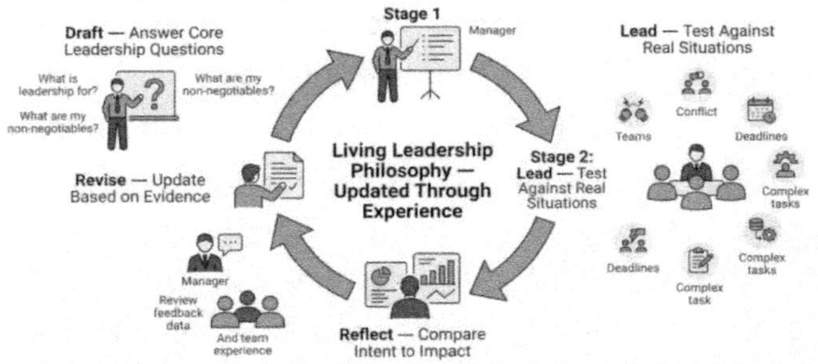

Cultivating a dynamic leadership philosophy involves constant application and refinement.

3.8 Using AI-Powered Assessments and Personality Tools

The landscape of AI-powered leadership assessment has advanced significantly in the last several years, and the tools available to individual managers—not just to enterprise learning and development organizations—now include capabilities that were previously accessible only through expensive executive coaching engagements. Understanding what these tools can and cannot reliably offer, how to use their outputs effectively, and where their limitations require you to supplement them with human judgment is a practically important form of management literacy that this section addresses directly.

Personality and behavioral style assessments—tools built on frameworks including the Big Five personality dimensions, the DISC behavioral model, the Myers-Briggs Type Indicator, the Enneagram, the Hogan Leadership Forecast series, and others—are most useful as starting points for self-directed inquiry rather than as definitive characterizations. The research on the predictive validity of these instruments varies considerably: some have stronger empirical foundations than others, and all of them simplify the genuine complexity of human personality for the practical purpose of making relevant patterns legible. The manager who uses assessment results to identify behavioral tendencies that warrant examination and testing—rather than as explanations that excuse or justify them—is using these tools correctly. 'My assessment says I am highly directive' is a useful hypothesis to test against your team's experience. It is not a license to remain highly directive regardless of the situation.

AI-powered coaching platforms that integrate with assessment instruments provide a continuous development experience that is qualitatively different from a single assessment event. These platforms use the assessment baseline as a foundation, then layer in ongoing reflective prompts, behavioral observation checkpoints, and specific micro-practice recommendations calibrated to the individual manager's development targets. Some platforms integrate with calendar and communication data (with appropriate consent) to provide contextual coaching based on upcoming meetings, recent interactions, or observed communication patterns. The quality of these platforms varies significantly, and the manager who approaches them as one input among several—rather than as an authoritative development prescription—will get the most value from them.

Three-hundred-sixty-degree feedback tools deserve particular attention because they address the most significant structural limitation of self-directed development: the inability to observe your own behavior as others experience it directly. A well-implemented three-sixty process—one that collects feedback from a meaningful number of respondents using behaviorally specific questions, ensures genuine anonymity, and provides synthesis that identifies patterns rather than just aggregating scores—provides management data that has no equally reliable substitute. The AI enhancements to modern three-sixty platforms include natural language processing to extract themes from open-ended responses, sentiment analysis to identify the emotional valence of qualitative feedback, and comparative analytics to benchmark a manager's feedback profile against relevant peer groups. These enhancements make it easier to identify the signal in what can be an overwhelming volume of feedback data and to translate that signal into a specific, actionable development priority rather than a diffuse aspiration to 'improve communication' or 'be more decisive.'

Diagram 2.8: AI Assessment Tools — Manager's Decision Guide

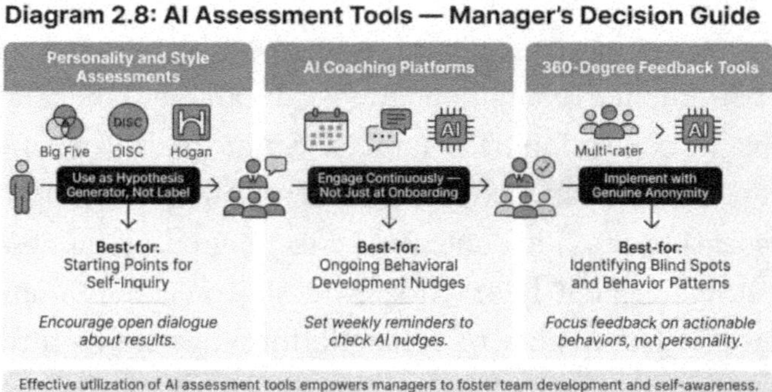

3.9 Manager's Checklist — Building Your Leadership Foundation

The practices below are not one-time activities. Each represents an ongoing investment that builds value over time as it becomes habitual and as you develop greater precision in applying it. Revisit this checklist quarterly and use it to identify areas where your foundation-building effort has been most and least consistent.

3.9.1 Self-Awareness and Emotional Intelligence

- Complete at least one formal personality or behavioral style assessment within your first sixty days and engage with the results not as a verdict but as a set of hypotheses to test against your actual leadership behavior over the following months.

- Identify your most reliable trigger—the type of situation, interaction, or outcome that most reliably activates a defensive or reactive emotional response— and develop a personal protocol for recognizing and managing it before it shapes your visible behavior.

- Implement a brief end-of-week self-reflection practice: identify one leadership interaction that went better than expected and understand specifically why; identify one that went worse than expected and identify specifically what you would do differently.

- Ask for feedback from at least one team member each month with a specific, behaviorally focused question: not 'how am I doing?' but 'in the last month, what is one specific thing I did as a manager that made your work harder or easier than it needed to be?'

- Track your emotional state at the start of key management interactions—meetings, one-on-ones, performance conversations—and notice how your state correlates with the quality of the interaction. Develop a preparation ritual that helps you reach your most productive state before high-stakes conversations.

3.9.2 Trust and Psychological Safety

- After each significant decision, communicate not just the decision but the reasoning: what information you used, what alternatives you considered, why you concluded as you did, and what would cause you to revisit the decision.

- When a team member makes an error or misses a target, conduct a structured learning conversation rather than a blame-oriented review: what happened, what contributed to it, what we can change to prevent it, and what the team member needs to develop the capability to handle it differently.

- Actively solicit contrary perspectives in team discussions and decisions. When consensus appears quickly, deliberately ask for the strongest argument against the position that seems to be winning.

- Identify one area of genuine uncertainty in your management practice and share it transparently with your team: 'I am still learning how to handle X well—

here is my current approach, and I want to improve it. What have you observed?'

- Conduct a psychological safety pulse check with your team at least once per quarter using a structured conversation or anonymous survey that assesses whether team members feel safe to speak up, admit mistakes, and challenge direction without fear of negative consequences.

3.9.3 Leadership Development and AI Tools

- Implement a structured three-hundred-sixty-degree feedback process within your first ninety days, ensuring genuine anonymity, using behaviorally specific questions, and engaging with the results in a development-oriented rather than defensive mode.

- Identify an AI-powered leadership coaching or assessment platform that aligns with your development priorities and engage with it for at least thirty minutes per week over a ninety-day pilot period.

- Write a first draft of your personal leadership philosophy within sixty days—no more than two pages, focused on your foundational beliefs about leadership rather than your aspirational traits—and share it with your team as a basis for accountability.

- Establish a regular leadership development peer group—two or three fellow managers with whom you meet monthly to discuss leadership challenges, share feedback, and hold each other accountable to stated development priorities.

- Identify one specific leadership competency you will focus on developing over the next ninety days, define two or three specific behavioral practices that represent that competency in action, and track your application of those practices weekly.

3.10 What to Carry Forward

The leadership foundation you build in your first year of management will determine not just the performance of your team today but the trajectory of your entire management career. Self-awareness, emotional intelligence, the capacity to build trust and psychological safety, and a clear, examined personal leadership philosophy are not qualities that appear fully formed through the effort of will. They are built incrementally, through the daily practice of attention, feedback, reflection, and deliberate behavioral adjustment. The investment required is real. The return on that investment—in team performance, in personal career sustainability, in the quality of the relationships you build over a management career—is among the highest available to any professional.

AI-powered assessment and feedback tools have genuinely expanded what is available to managers who want to deliberately build their foundation. The continuous feedback loops, the pattern recognition across large volumes of qualitative data, and the personalized development recommendations that modern platforms provide were not accessible to previous generations of new managers. Use them as inputs that sharpen your self-knowledge, not as authorities that define your capability. The fundamental work—the honest examination of your behavior, the willingness to change what the evidence says needs changing, the sustained commitment to the values and principles you have articulated as your foundation—is work that only you can do.

Every effective manager you will meet across your career has, at some point, sat in a meeting like Jordan's—confronted with a gap between who they believed themselves to be and who they actually were in the experience of the people they led. The best of them got curious rather than defensive. They asked questions rather than offering explanations. They revised their self-understanding and their behavior based on what they learned. And they emerged from that process with a leadership foundation that was more honest, more grounded, and more genuinely effective than the one they walked in with. That process is available to you, right now. It is one of the better investments you will ever make.

4 Communicating with Clarity and Impact

4.1 The Message That Missed

Marcus had just finished what he considered one of his better all-hands updates—a tight twenty-minute overview of the quarter's priorities, a summary of the organizational changes that were coming, and a confident close about the team's position heading into the second half of the year. He had prepared carefully, covered every major point on his list, and kept the message clear and concise. Two days later, his manager mentioned casually that three of Marcus's team members had separately approached her with questions about whether the upcoming organizational changes meant their roles were at risk. Marcus was floored. He had specifically addressed his team's stability. He had been explicit. He went back to his notes and read exactly what he had said: 'The organizational structure will evolve over the next two quarters, and I expect the team to navigate this well.' He could not have been clearer. What he could not see was that 'will evolve' and 'navigate this well' are the precise kind of hedged, unanchored language that trained communication professionals use to avoid making commitments, and that anxious team members hear as confirmation of exactly what they fear most.

This scenario illustrates a principle that distinguishes excellent managers from merely competent ones: communication is not what you say, but how you say it. It is what your audience understands. The gap between those two things—often invisible to the sender, always consequential for the team—is where most management communication fails. And it fails not because managers are careless or dishonest, but because effective communication requires skills and habits that are rarely taught explicitly, that must be rebuilt for the specific demands of a management role, and that are complicated in new and interesting ways by the tools, formats, and expectations of the modern workplace.

The good news is that communication is among the most learnable of management competencies. The frameworks are well-tested. The practices are specific and behaviorally concrete. And the AI tools now available to support communication—writing assistants that help with clarity and tone, transcription and summarization tools that make the substance of important conversations permanently accessible, and analytics tools that surface patterns in how communication is landing—have meaningfully expanded what individual managers can do without a professional communications team. This chapter provides the framework, the practices, and the AI integration strategy to make you a manager whose communication consistently informs, aligns, and motivates.

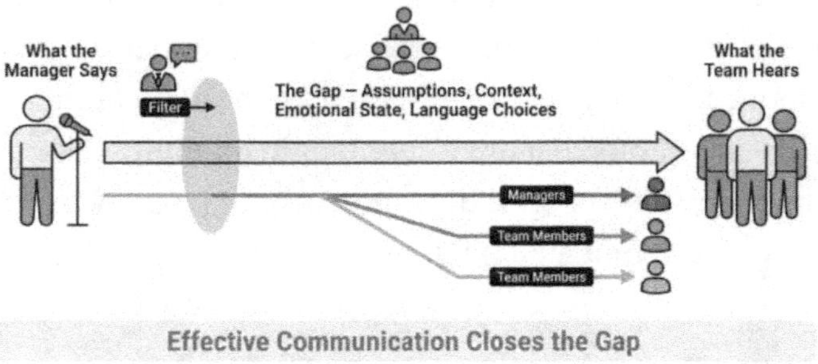

4.2 Why Manager Communication Is Different

Management communication operates under a set of constraints and responsibilities that fundamentally distinguish it from the communication required in individual contributor roles, and understanding those distinctions is a prerequisite for building the right skills. The first constraint is positional asymmetry: when you communicate as a manager, you are not just one participant in a professional conversation. You are the person with organizational authority over the people you are addressing. This asymmetry shapes how everything you say is received, often in ways that are invisible to you and entirely predictable from the listener's perspective. A casual observation from a manager is heard as a priority signal. An offhand question is experienced as an evaluation. A moment of visible frustration is read as a signal about the safety of bringing problems forward. The manager who is not aware of this asymmetry communicates constantly and inaccurately, sending signals they did not intend and missing the reactions those signals produce because they are not looking for them.

The second constraint is organizational consequence: the information managers communicate has a direct, material impact on the people who receive it. When Marcus said The organizational structure will evolve,' the consequence of that ambiguity was not a mildly confused conversation. It was three team members spending two days in escalating anxiety, interpreting every organizational signal through the lens of that ambiguity, and eventually taking their fears to Marcus's manager—a sequence that damaged trust, consumed organizational bandwidth, and required a corrective communication conversation that took considerably more time and relational capital than clarity in the original message would have cost. Every significant communication a manager delivers has a cost of ambiguity and a benefit of clarity that is worth calculating explicitly before the message goes out.

The third constraint is audience complexity: managers communicate simultaneously, often with a single message, to people with different information contexts, organizational concerns, roles, and emotional relationships with the content. The senior team member who has been through three organizational restructurings hears 'organizational evolution' differently than the team member who has been in their role for eight months and has never experienced significant change. The team member whose project is most directly affected by a priority shift hears a reprioritization announcement differently than the team member whose work is relatively unaffected. The manager who communicates as if the audience were a homogeneous group with shared context, shared concerns, and shared emotional responses will consistently leave a significant portion of the audience confused, anxious, or alienated—even when the message itself is technically accurate.

AI tools are changing the practice of management communication in ways that are both practically significant and analytically interesting. Writing assistance tools that help managers review their drafts for clarity, tone, and potential misinterpretation are available through platforms that most managers already use daily. Transcription and summarization tools that make meeting content permanently accessible, searchable, and actionable have dramatically reduced the overhead cost of distributed and asynchronous communication. Communication analytics tools that surface patterns in email response rates, meeting participation, and team sentiment provide managers with data on how their communication lands, previously available only through direct observation and informal feedback. Used deliberately, these tools can meaningfully improve the precision, reach, and consistency of manager communication. Used thoughtlessly, they can accelerate the distribution of poorly designed messages to more people, faster. The difference is in the design of the communication itself—and that is the human work that this chapter is about.

4.3 The Manager's Communication Toolkit

Effective management communication rests on three foundational capabilities that must be developed distinctly and practiced deliberately: listening, speaking, and writing.

Most managers arrive in the role with some level of competence in all three, developed through professional experience that valued individual output. Management requires these competencies at a qualitatively different level and with a qualitatively different orientation: less about producing communication efficiently and more about producing the specific understanding and response that the situation requires.

4.3.1 Listening as a Management Skill

Listening is the most undervalued and most consistently underdeveloped of the three foundational communication capabilities, and it is also the one with the most direct impact on a manager's ability to lead effectively. The information a manager needs to make good decisions—about team dynamics, individual performance, organizational obstacles, emerging risks, and what is actually happening in the work rather than what is being reported about it—lives primarily in the minds and experiences of the team members. That information surfaces only when the team member believes the manager will genuinely receive it, will not react defensively to it, and will act on it appropriately. Creating those conditions requires a quality of listening that most professional environments neither teach nor reward: active, curious, and suspensive of the manager's own agenda.

Active listening in a management context is a set of specific behaviors, not a state of mind. It includes physical and virtual signaling that you are fully present: eye contact and body language that communicate engagement in in-person settings, video presence, and reduced distraction signals in virtual settings. It includes verbal acknowledgment—reflective statements and summaries that demonstrate that what was said was actually heard, rather than processed through the filter of what you expected or wanted to hear. It includes questioning that deepens understanding rather than narrowing it: questions that open up the topic rather than confirming a predetermined interpretation. And it includes the discipline of tolerating silence—allowing the pauses in which team members formulate their actual thoughts rather than their expedient first responses.

The quality of your listening directly determines the quality of the information you receive, which directly determines the quality of the decisions you make. The manager whose team members know from experience that raising a concern will be genuinely heard and thoughtfully addressed receives more concerns earlier—before they become crises. The manager whose team members have learned that raising concerns results in defensive explanations or dismissals receives information only when it is impossible to withhold it, i.e., when the crisis has already arrived. Building a reputation as a manager who genuinely listens is not a soft relational achievement. It is an operational intelligence asset.

Cultivating active listening habits within leadership teams builds a culture of trust and operational excellence.

4.3.2 Speaking with Precision and Purpose

Spoken communication as a manager operates in several distinct registers, each with its own requirements: the informal daily interactions that set the relational and operational tone of the team; the structured one-on-one conversations that are the primary vehicle for coaching, feedback, and individual alignment; the team meetings that create shared understanding and collective alignment on priorities and progress; and the high-stakes conversations—performance discussions, change announcements, difficult feedback—that have outsized and lasting impact on the individuals involved and on the team's psychological environment. Becoming a strong spoken communicator as a manager requires developing specific skills for each register rather than applying a single approach across all contexts.

The highest-leverage spoken communication skill for most new managers is precision in language about expectations and priorities. Vague language—' let's keep an eye on this,' 'I expect us to do better,' 'we need to be more strategic about this'—does not communicate an expectation.

It conveys the shadow of an expectation, which people interpret through their own assumptions and concerns. Specific language—' I want a weekly status update on this project every Friday by noon,' 'the quality target for next month is X and here is specifically what that means,' 'when I say strategic, I mean that every decision about resource allocation should be evaluated against these three criteria'— gives team members something they can act on, align to, and report progress against. The specificity cost is a few additional seconds per communication. The clarity benefit is enormous.

The second highest-leverage spoken communication skill is the explicit checking of understanding—not 'does that make sense?' which is a yes-or-no question that reliably produces 'yes' regardless of actual comprehension, but the more demanding and more honest 'can you walk me through what this means for your work this week?' or 'what are the implications of this decision for the project you are currently working on?' These questions reveal understanding rather than assuming it, and they give the manager actionable information about where additional clarity is needed before the gap between what was said and what was understood becomes operational.

Tone is the dimension of spoken communication that most managers underinvest in as a development focus, partly because it is the most difficult to observe in yourself and partly because the organizational culture of most professional environments does not provide explicit feedback on it. Tone is not just pleasantness or unpleasantness—it is the emotional signal carried by how you say things, which often has more impact than what you say. The manager who delivers accurate, helpful feedback in a tone that reads as contemptuous or dismissive has failed to communicate, regardless of the content's accuracy. The manager who delivers a difficult organizational message in a tone that reads as genuine and empathetic—that signals that the difficulty is acknowledged and that the people affected matter—creates a qualitatively different response than the same message delivered in a flat, transactional register. Developing awareness of your tonal range and intentionality about which register you are using in which context is a communication development priority that pays compounding dividends.

Diagram 3.3: The Spoken Communication Registers for Managers

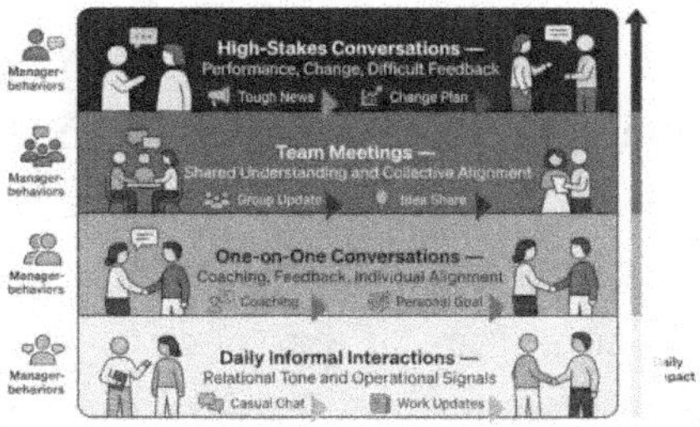

Effective communication requires managers to deliberately adjust their register to match the situation.

4.3.3 Writing That Informs, Aligns, and Motivates

Written communication is the primary operational medium of modern management. Email, messaging platforms, project management tools, documents, presentations, and the asynchronous threads that sustain distributed team work are all written mediums, and the quality of a manager's writing—its clarity, structure, appropriate level of detail, and effectiveness in producing the intended response—has a direct impact on team efficiency, alignment, and trust. Poor written communication is not just inefficient; it is also harmful. It creates the compounding overhead of clarification conversations, misaligned work, and the kind of accumulated misunderstanding that gradually erodes team trust in ways that are difficult to trace back to their source.

The most important principle of effective management writing is that every piece of written communication should be designed for a specific reader, with a specific purpose, to produce a specific response. This principle sounds obvious and is rarely applied. The typical management email is written to transfer information that the manager needs to transmit, without explicit design for who will receive it, what they need to understand and act on, and what structural and language choices will most efficiently and reliably produce that understanding and action. Writing designed for the reader's needs rather than the writer's mental model requires a brief but genuine act of perspective-taking before each significant communication: who is reading this, what do they already know, what do they need to know to act on this effectively, and what is the specific action or outcome I need this communication to produce?

Structure is the most powerful tool available for improving the clarity and effectiveness of management writing, and it is the most underused. A well-structured written communication—one that leads with the most important information (bottom line up front), provides organized supporting detail in logical sequence, and closes with explicit next steps or calls to action—can be understood and acted on in a fraction of the time of an unstructured communication containing the same information. The principles of document structure apply across formats: the email that leads with the key ask before providing context; the project update that leads with the current status and critical decisions needed before providing detail on specific work streams; the change announcement that leads with what is changing and why it matters before explaining the implementation timeline and process.

AI writing assistants offer substantive value for management communication when used as editors rather than ghostwriters. A writing assistant that helps you review a draft for clarity of structure, identifies language that is likely to be ambiguous or misinterpreted, suggests alternative phrasings that improve precision or tone, and flags missing context that the reader may need is an extremely useful tool for producing communication that achieves its purpose more reliably than unassisted drafting. A writing assistant used to generate complete communications that the manager sends without meaningful revision is a tool being used in a way that bypasses rather than enhances the manager's judgment, and that produces communication that is accurate in a generic sense but lacking the contextual, relational, and organizational specificity that makes management communication genuinely effective. The manager's voice, knowledge of the specific team and situation, and judgment about what this particular audience needs—these are what make the communication work. AI assistance should sharpen those inputs, not replace them.

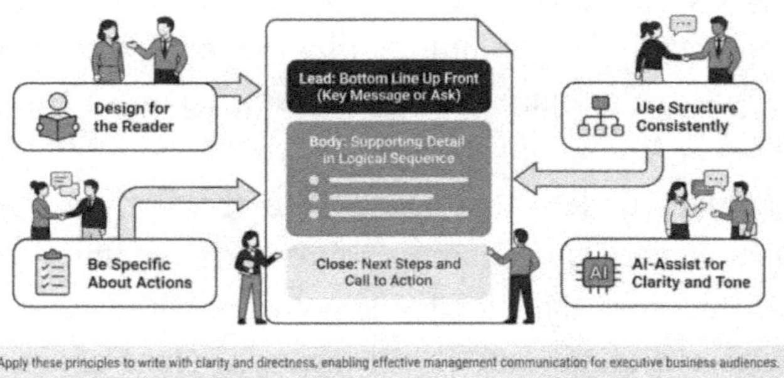

4.4 Running Effective Meetings That People Actually Value

Meetings are the primary mechanism through which managers create collective alignment, make shared decisions, and build the team relationships that sustain collaboration over time. They are also the primary mechanism through which organizational time is wasted, motivation is eroded, and the implicit message is sent that the manager does not respect the team's time and attention as finite and valuable resources. The difference between these two outcomes is almost entirely determined by design: whether the manager approaches meetings as a purpose-driven tool to be used deliberately and shaped to produce the specific outcome it is meant to achieve, or as a default activity that fills calendar space and provides a simulacrum of coordination without substance.

Every meeting your team attends has a real cost measured in the aggregate time of all participants. A one-hour meeting with six people costs six hours of organizational time, regardless of whether those six people could have achieved the same outcome through a twenty-minute asynchronous exchange. Making that cost explicit—even just privately, as part of your own meeting design practice—is one of the most immediately useful disciplines a new manager can develop. It creates a natural incentive to invest in meeting design: to define the objective clearly before scheduling, to invite only the people whose presence is genuinely necessary for that objective, to structure the agenda so that the time is used for the discussion and decision-making that requires synchronous presence rather than for status updates that could have been read in advance.

The taxonomy of management meetings breaks into four distinct types that serve different purposes and require different design: the decision meeting, structured to reach a specific decision with the people who have the information and authority needed; the problem-solving meeting, structured to diagnose a complex challenge and develop response options; the coordination meeting, structured to ensure alignment on current work, upcoming priorities, and cross-functional dependencies; and the communication meeting, structured to deliver significant information and allow for questions and discussion. Each type has an optimal structure, an appropriate duration, a specific set of pre-work requirements, and a clear success criterion. Conflating these types—asking a decision meeting to also serve as a status update and a brainstorming session—is one of the most common and most consequential meeting design errors managers make.

The facilitation skills that make meetings genuinely productive are distinct from the subject matter expertise that makes a manager technically credible, and they deserve deliberate development. Effective meeting facilitation includes: opening the meeting with a clear statement of objective and agenda so that participants know what a successful outcome looks like; managing participation so that the discussion does not default to the most assertive voices at the expense of important but quieter perspectives; keeping the discussion on purpose rather than allowing productive tangents to consume the time available for the meeting's primary objective; synthesizing and confirming agreement before moving from one decision or topic to the next; and closing the meeting with an explicit summary of decisions made, actions committed to, and owners and timelines established for each. These behaviors are entirely learnable and have an immediate, visible impact on the quality and efficiency of meeting outcomes.

Diagram 3.5: The Four Meeting Types — Purpose and Design

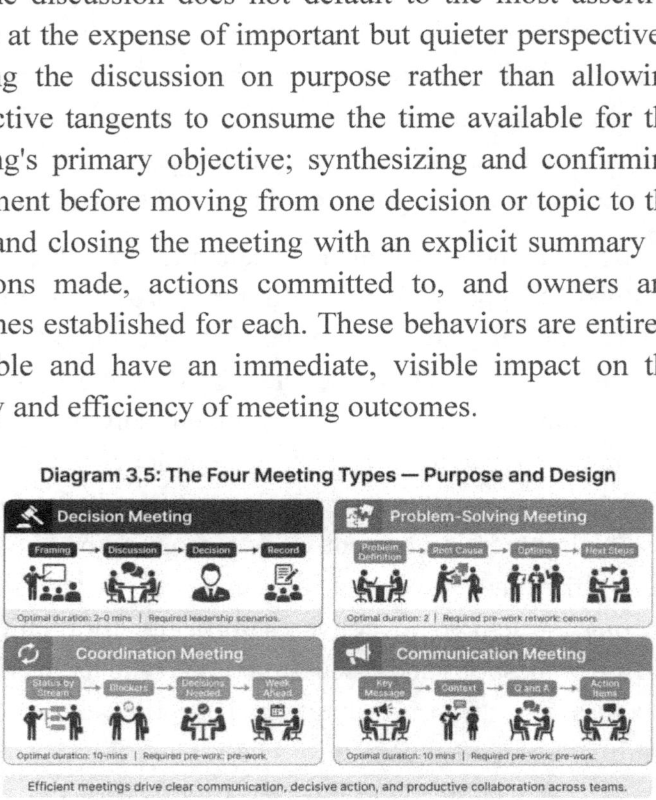

4.4.1 AI Tools for Meeting Effectiveness

AI-powered meeting tools have advanced sufficiently to serve as a genuine productivity lever for managers who deliberately integrate them into their meeting cadence. The most mature and broadly useful category is AI transcription and summarization: tools that join meetings as a participant (with team disclosure and consent), generate a real-time written transcript, and then produce structured summaries that identify key decisions, action items with owners and timelines, and the main discussion points. These tools address one of the most persistent and costly inefficiencies in management practice: the gap between what is discussed and decided in meetings and what is actually acted on afterward, which is largely a function of the imperfect, selective memory that manual note-taking produces.

The practical value of AI meeting summarization extends beyond simply replacing manual notes. AI-generated transcripts make meeting content searchable and permanently accessible, so the team member who missed the meeting has genuine access to what was discussed rather than a summary filtered through the note taker's interpretation. It means that follow-up discussions about what was decided in a prior meeting can be resolved by reference to the actual transcript rather than competing recollections. It means that patterns across meetings—recurring topics that never reach resolution, decisions that are revisited because action items were not followed up, discussion patterns that consistently exclude certain team members—become visible in data rather than invisible in memory. These are organizational intelligence assets that accumulate over time and that make the team progressively more effective at the coordination function that meetings serve.

AI scheduling and calendar optimization tools offer a second significant value category for meeting management.

Intelligent scheduling assistants can identify optimal meeting times based on participant availability and preferred working hours, suggest meeting durations based on the agenda and objective type, flag meetings that appear to lack clear objectives or required participants, and help managers audit their calendar periodically to identify meeting patterns that are consuming disproportionate organizational time without commensurate value. These tools do not make the design decisions that determine whether a meeting is worth holding—that remains the manager's judgment—but they reduce the coordination overhead of scheduling and provide useful data for the periodic calendar audit that most managers know they should conduct but rarely do.

4.5 Communicating Up, Down, and Across the Organization

Management communication happens in three directions simultaneously, each with its own requirements, challenges, and success criteria. Communicating downward to your team—the most visible and most practiced dimension—is the focus of most management communication training. But communicating upward to your manager and senior leadership, and laterally across to peers and cross-functional partners, are equally important and significantly more complex dimensions that new managers routinely underinvest in.

Communicating upward effectively requires translating the operational reality of your team's work into the strategic and organizational language that senior leaders use to evaluate priorities, allocate resources, and make decisions.

Senior leaders are not primarily interested in the technical details of how the work is done; they are interested in whether the work is on track to achieve the stated business outcomes, whether there are risks that require organizational attention or resources, and whether there are insights from the team's experience that should inform strategic decisions. The manager who provides upward communication in operational detail without strategic translation is making their manager do work that should be done at the team level. The manager who provides strategic translation without sufficient operational grounding to support the claims being made is providing communication that cannot withstand scrutiny. The skill is calibrating the level of detail and abstraction to the specific audience and decision context.

Communicating laterally—with peers in other functions whose teams interact with yours—is a dimension that many new managers neglect because it lacks the clear hierarchical structure that governs upward and downward communication. There is no organizational mandate that requires you to keep your cross-functional partners informed or to proactively manage the relationship between your team's work and theirs. But the absence of that mandate does not mean the absence of consequence for failing to do it. Teams whose managers maintain strong lateral communication relationships have faster access to information that matters, more collaborative resolution of cross-functional conflicts, more visibility into organizational priorities and shifts that affect their work, and more organizational goodwill to draw on when they need cooperation from other teams under conditions of organizational scarcity. Building strong lateral relationships is an investment in your team's operating environment that pays forward in ways that are difficult to quantify and impossible to replace once they have been neglected long enough to erode.

Both upward and lateral communication benefit significantly from the consistency and professionalism that AI writing assistance can support. The senior leadership update, drafted quickly and unreviewed, reflects poorly on the manager and the team. The same update, drafted quickly, reviewed by an AI writing assistant for structure, clarity, and an appropriate level of detail, and revised before sending, is a meaningfully different communication product that requires only marginally more time to produce. Making AI-assisted review a standard step in the preparation of any significant upward or lateral communication—rather than a special effort reserved for high-stakes occasions—is a workflow-level practice that compounds in reputational value over time.

Diagram 3.6: Communicating in Three Directions

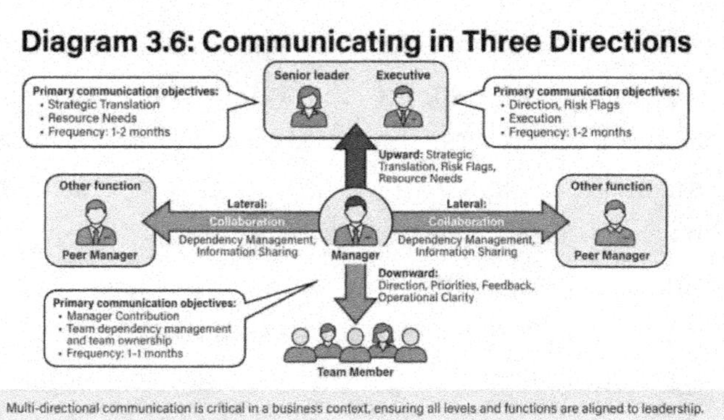

Multi-directional communication is critical in a business context, ensuring all levels and functions are aligned to leadership.

4.6 Delivering Difficult Messages with Empathy and Clarity

Difficult messages—performance feedback that requires honesty about significant gaps, organizational changes that affect team members' roles or security, unpopular decisions or those the team member disagrees with, or conversations about interpersonal behavior that must change—are among the most consequential communications a manager makes. They are also the ones most frequently avoided, diluted, or so wrapped in softening language that the core message fails to land with the clarity the situation requires. Understanding why difficult messages are difficult and developing a specific, repeatable approach to delivering them effectively are communication skills with direct and lasting impact on team performance and individual development.

Difficult messages are psychologically challenging to deliver for a specific reason: the manager cares, to some degree, about the relationship with the person they are communicating with, and delivering hard information creates the risk of damaging that relationship. This relational concern is not a weakness—it is evidence of the empathy that good management requires. The problem is that the strategies most managers employ to manage the relational risk—softening the message to the point of ambiguity, burying the difficult content beneath layers of positive framing, avoiding the message entirely until a crisis makes it unavoidable—are precisely the strategies that produce the worst outcomes both for the relationship and for the performance issue being addressed. The team member who does not receive clear, honest feedback about a significant performance gap cannot address it. The team member who receives ambiguous feedback has experienced relational discomfort without the developmental benefit. The team member who receives clear, honest, empathetic feedback has been respected as a capable adult who deserves the truth about their performance and the opportunity to address it.

A practical framework for difficult conversations that managers consistently find effective structures the interaction in three phases. The first phase is preparation: defining clearly, before the conversation begins, what specific behavior or outcome needs to change, what the impact of that behavior or outcome has been on the team or on the individual's performance, and what a successful outcome of the conversation looks like. Under-prepared, difficult conversations typically either default to excessive softening—because the manager has not committed clearly enough to the core message to deliver it unambiguously—or to excessive directness without context—because the manager has not thought through the why and the impact in a way that makes the feedback genuinely useful to the recipient.

The second phase is delivery: leading with the core message rather than burying it, using specific, behavioral language rather than evaluative or characterological language, and providing the impact and context that make the feedback meaningful rather than arbitrary. 'The last three project updates you submitted were submitted after the agreed deadline and included errors that required correction before distribution' is specific, behavioral, and actionable. 'You have been unreliable about your deadlines and your quality has been inconsistent' is evaluative, generalized, and defensive response producing. The distinction is not just linguistic. It is the difference between a feedback conversation that can produce genuine behavior change and one that produces a defensive argument about whether the characterization is fair.

The third phase is forward orientation: ensuring that the conversation ends not just with the problem clearly identified but with a specific, agreed-upon path forward. What behavior is expected going forward? What support is available? What will the manager do differently to support the change? What is the timeline for reviewing progress? Difficult conversations that end with clear problem identification but no path forward leave the team member with clarity about the negative assessment and no clarity about what to do with it. The path forward is what makes the conversation developmental rather than merely evaluative, and it is the phase that most managers, having navigated the discomfort of the core message, rush through or omit entirely.

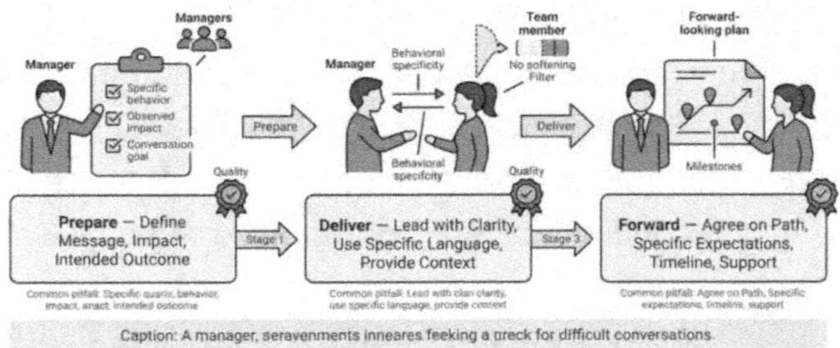

Caption: A manager, seravenments inneares feeking a breck for difficult conversations.

4.7 Mastering Async Communication for Hybrid and Remote Teams

Asynchronous communication—the exchange of information without the expectation of an immediate response—has become the primary operating medium for the majority of professional knowledge work, particularly for teams that span multiple time zones, work in hybrid arrangements, or include members with variable schedules or deep work commitments that make synchronous availability inefficient. Most managers have not deliberately designed their approach to async communication; they have adapted the habits of in-person, synchronous communication to digital mediums without the structural changes that would make those habits effective in the async context. The result is the characteristic dysfunction of most hybrid and distributed team communication: overwhelming message volumes that require constant monitoring, critical information buried in unstructured threads, the inequity between team members whose working hours overlap with the majority and those whose working hours do not, and the ambient anxiety of teams that are never quite sure whether they are missing something important.

Designing effective async communication requires explicit decisions about three dimensions: channels, norms, and structures. Channel design is the decision about which communication medium serves which communication purpose. Using a real-time messaging platform for information that requires no immediate response creates false urgency and the cognitive burden of constant monitoring. Using email for collaborative discussion that benefits from threading, reactions, and informal engagement is inefficient and loses the conversational quality that discussion requires. Using a project management tool for interpersonal coordination that is not task-specific adds organizational overhead without adding clarity. The most effective async communication systems use a small number of carefully chosen channels, with explicit norms about what types of communication belong in each, and the discipline to route communications to the correct channel rather than defaulting to whatever channel requires the least thought.

Norm design is the set of explicit agreements that define how the team uses its communication channels: expected response time for different message types and channels, what topics warrant interruption versus async response, how documents and decisions are organized and stored, how meeting artifacts (transcripts, summaries, decisions) are distributed and where they live, and what communication behaviors represent poor digital citizenship—excessive message volume, inadequate context provision, inappropriate urgency signaling. The most effective communication norms are developed collaboratively with the team rather than imposed by the manager, because team members have direct operational insight into what is and is not working in their communication environment, and because norms developed collaboratively are consistently better followed than norms handed down from above.

Structure design is the specific formatting and organizational standards that make async communication self-sufficient—that allow a team member to read a message, document, or update without asking clarifying questions or attending a synchronous meeting to understand what is being said and what is needed from them. The well-structured async communication leads with what the reader needs to know and do, provides organized supporting context in a scannable format, clearly labels any time-sensitive elements, and ends with explicit statements of who is responsible for what action by when. These are simple disciplines that require modest additional effort per communication and produce enormous collective efficiency across a team receiving hundreds of such communications per week.

Diagram 3.8: The Async Communication Design Framework

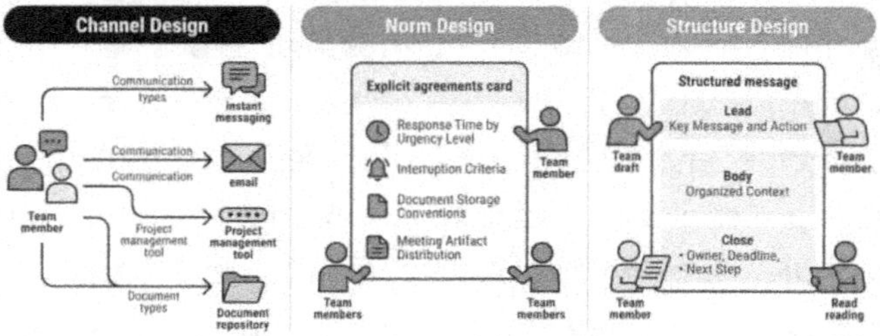

This three-dimension framework optimizes asynchronous communication by defining **clear channels, behavioral norms,** and **structured message templates.**

4.8 Building a Communication Cadence
That Keeps Teams Aligned

A communication cadence is the regular, structured rhythm of information exchange that maintains team alignment without requiring constant ad hoc coordination. It is the design decision that replaces the question 'how do I make sure my team stays aligned?' with a set of recurring, predictable communication events that each serve a specific alignment purpose and that collectively ensure that the team is never more than a short time away from the next opportunity to surface issues, confirm priorities, make decisions, and connect as a group. Cadence is the infrastructure of operational clarity.

The standard management communication cadence includes three primary rhythms at different time horizons. The weekly rhythm is the operational heartbeat: the regular team meeting that ensures collective alignment on current week priorities, the review and reset of any decisions or priorities that have shifted since the prior week, and the surfacing of blockers and cross-team dependencies that require coordinated resolution. The weekly rhythm also includes the individual one-on-ones, which are the primary vehicle for personalized coaching, feedback, development conversations, and the kind of direct, unfiltered individual communication that team settings do not allow. The monthly rhythm is the strategic and developmental pulse: the team retrospective or reflection session that examines what the team is learning about how it works and how it could work better; the individual development conversation that takes the longer view of each team member's growth, career trajectory, and current engagement; and the upward and lateral communication that keeps the manager's manager and cross-functional partners current on the team's direction and progress.

The quarterly and annual rhythms address the strategic and planning horizons: the goal-setting and strategy alignment sessions that connect team priorities to organizational direction, the formal performance review processes that provide team members with structured feedback on their development and contribution, and the deeper planning conversations that anticipate the next period's challenges and resource requirements. Managers who operate with all three rhythms running simultaneously have a team that is consistently better informed, more aligned, and more capable of sustained high performance than teams whose communication is driven by urgency and event rather than by deliberate cadence design.

The practical implementation of a communication cadence requires discipline to protect recurring events from the inevitable pressure of short-term urgency that fills management calendars. Weekly one-on-ones, in particular, are the first to get canceled when schedules become crowded—and they are the most irreplaceable element of the individual management relationship. The team member whose one-on-one is consistently rescheduled or canceled receives an unambiguous signal about their relative priority in the manager's attention and investment, regardless of any stated commitment to their development. Protecting the cadence under pressure is one of the most visible ways a manager demonstrates that their stated priorities align with their actual behavior. This consistency is foundational to the trust that the cadence exists to support.

Diagram 3.9: The Manager's Communication Cadence

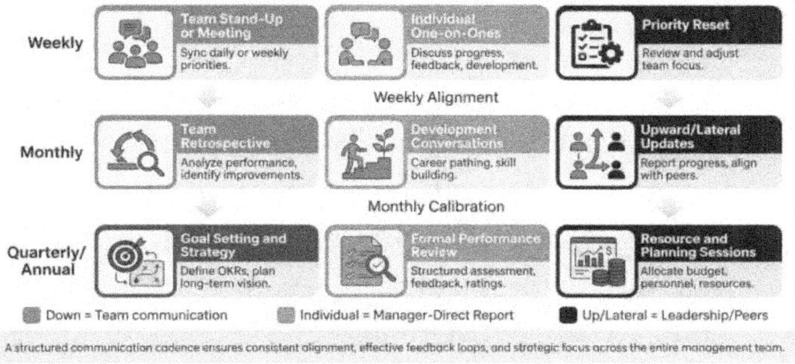

4.9 Manager's Checklist — Communicating with Clarity and Impact

Communication competency is built through consistent practice and deliberate feedback. Use this checklist to identify the highest-leverage communication improvements available in your current context and track your development over a ninety-day horizon.

4.9.1 Core Communication Practices

- Before sending any significant communication—email, document, meeting agenda—write one sentence defining who is reading it, what they need to understand, and what specific action or response you need the communication to produce. If you cannot write that sentence clearly, the communication needs more design before it goes out.

- Audit your last ten emails for structure: do they lead with the most important information, organize supporting detail logically, and close with explicit next steps? Identify the most common structural weakness in your writing, and create a personal template or checklist to address it.

- Review your calendar for the past two weeks and classify each meeting by type (decision, problem-solving, coordination, communication). Identify meetings that conflated multiple types or that lacked a clear primary objective. Design a brief pre-meeting template that defines the objective, required participants, pre-work, and success criteria for your ongoing meetings.

- Identify the three most important upward communications you will make in the next ninety days and practice translating the operational content of each into the strategic language your manager or senior leadership uses to evaluate the work.

- For the next thirty days, end every significant meeting with an explicit verbal summary of decisions made, actions committed to, owners identified, and timelines confirmed. Note whether this practice changes the clarity and follow-through of meeting outcomes.

4.9.2 Difficult Conversations and Feedback

- Identify the difficult conversation you have been postponing and prepare for it using the three-phase framework: what is the specific behavior or outcome, what is its impact, what is the path forward? Schedule it within the next two weeks.

- After each feedback conversation you conduct, write a brief self-assessment: did the recipient understand the specific behavior that needs to change? Did the conversation end with a clear, agreed-upon path forward? What would you do differently?

- Practice the precision distinction in language: replace three evaluative or characterological feedback statements you have used recently ('unreliable,' 'not strategic,' 'difficult to work with') with specific, behavioral equivalents that describe observable actions and their observable impact.

- Establish a personal norm for checking understanding at the end of significant communications: ask a question that requires the listener to demonstrate understanding rather than confirm it.

- Identify one team member who may benefit from a proactive, direct conversation about a pattern you have observed in their work or behavior—not a crisis conversation, but a developmental investment—and schedule it before the end of the current month.

4.9.3 AI Tools for Communication

- Pilot an AI writing assistant tool for one category of management communication—weekly team updates, one-on-one preparation notes, or upward reporting— for thirty days. Track whether the quality of that communication improves, as measured by the clarity of responses and the frequency of clarifying questions it generates.

- Implement an AI meeting transcription and summarization tool for your weekly team meeting for one month, with team disclosure and consent. Evaluate whether the quality of action item follow-through improves when team members have access to accurate, searchable meeting records.

- Use AI writing assistance to review your last three significant written communications and identify the most common feedback themes: clarity, structure, tone, and specificity. Use those themes as input for a focused 30-day writing improvement practice.

- Establish a personal AI use protocol for communication: define which communication tasks you will use AI assistance for, what your review standard is for AI-assisted outputs before sending, and how you will ensure that AI assistance supports rather than replaces your authentic voice and contextual judgment.

- Evaluate the communication channels your team currently uses and identify one channel that is being used for the wrong purpose or without adequate norms. Draft a one-page channel norm document for that channel, share it with the team for input, and implement the agreed norms for a sixty-day trial period.

4.10   What to Carry Forward

Communication is the medium through which management happens. Every other leadership practice—

direction-setting, coaching, decision-making, culture-building—is expressed and delivered through the quality of the accompanying communication. A brilliant strategy communicated ambiguously leads to confused execution. An excellent coaching insight delivered without empathy and behavioral specificity produces defensiveness rather than development. A critical organizational change announced without transparency and honest engagement with the concerns it raises produces anxiety and eroded trust. The investment in becoming a genuinely excellent manager communicator is not separate from the investment in becoming a genuinely excellent manager. They are the same investment.

The AI tools available to support management communication represent a practical opportunity that deserves to be treated as such—not as a shortcut that substitutes for the work of becoming a better communicator, but as an assistance system that helps you communicate more precisely, more consistently, and more effectively than your unassisted effort alone would produce. The writing assistant that helps you catch ambiguous phrasing before it leads to a team-wide misunderstanding. The meeting summarization tool that makes the decisions from Tuesday's meeting permanently accessible, rather than remembered imperfectly by six people who were all partially distracted. The communication analytics that surface the pattern of response rates and follow-through, telling you something concrete about how your communication is landing. These tools work best in the hands of a manager who understands the principles of effective communication and uses AI assistance to apply them more consistently—not in the hands of a manager who is hoping that the tools will supply the principles.

Marcus, from the opening of this chapter, revised his communication approach after that conversation with his manager. He did not overhaul his entire style. He added two practices: he began every significant team communication with a brief self-audit against a question—' what will my most anxious team member hear in this message?'—and he ended every significant team meeting with a structured check for comprehension, asking specific rather than confirming questions. Six months later, he described those two practices as among the most impactful things he had done in his first year of management. The changes were small. The cumulative impact on his team's alignment, trust, and engagement was substantial. That is how communication development works: small, deliberate, consistent improvements that compound in operational and relational impact over time.

5 The Art and Science of Delegation

5.1 Scenario

It is 7:45 on a Tuesday morning, and Maya has already logged three hours of work. As a newly promoted engineering manager, she spent yesterday afternoon personally rewriting a junior developer's status report, jumped on a call to troubleshoot a deployment issue that her team lead could have handled, and drafted a project update email that, by her own admission, required one paragraph from her and four from people she manages. By the time her one-on-one calendar starts at ten, she is behind on the strategic roadmap review her director expects by Friday, and she has not once opened the workforce planning spreadsheet she promised to complete this week.

Maya's calendar is full, and her output looks high. But she is not managing — she is executing. The work she is doing belongs to her team, not to her. And the work that belongs to her — thinking, planning, prioritizing, developing people — is not getting done.

Her team, meanwhile, is quietly frustrated. They feel supervised instead of trusted. When Maya rewrites their reports without explanation, they stop trying to improve them. When she jumps into technical problems before they have a chance to be solved, she stops developing the problem-solving instincts that would make her a more valuable contributor. The team's capacity has not grown since Maya took over the role. Neither have they as professionals.

This scenario is not a management failure. It is a normal, predictable, and recoverable pattern — one that almost every new manager experiences in the first year. The solution is not to work harder. It is to understand delegation not as a task assignment trick, but as the primary mechanism through which a manager multiplies the capability of an entire team.

5.2 Why Delegation Is a Leadership Imperative

Delegation is the leverage point that separates managers who scale from managers who stall. When a manager holds onto work that someone else could and should own, that manager creates a bottleneck at the center of the team's operational flow. Every decision that requires their approval, every output they feel compelled to refine, and every problem they step in to solve is work that did not develop the person who should have done it, and time that did not go toward the strategic work only the manager can do.

The stakes extend beyond personal productivity. Teams where delegation is weak develop learned helplessness over time. Team members stop bringing their full capability to work because they have learned, through repeated experience, that the manager will step in before they finish. They stop taking intellectual risks because they know the output will be revised anyway. They stop growing into the next level of responsibility because they are never given the sustained ownership that growth requires. The long-term cost of under-delegation is measured in attrition, stagnation, and the quiet erosion of team capability that never shows up in any single week's metrics but accumulates into a serious talent problem over months.

Diagram 4.1: The Delegation Leverage Model

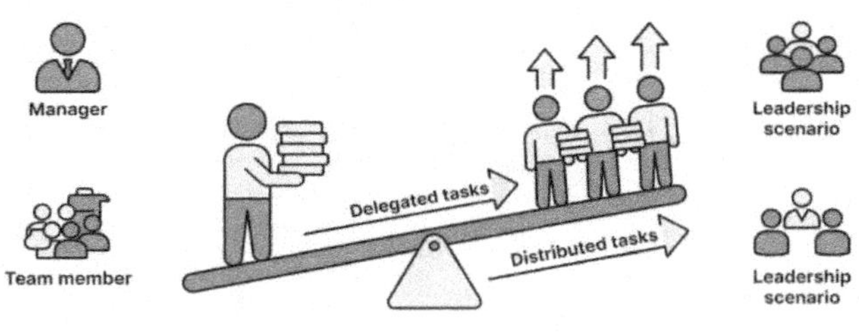

Effective delegation lifts team capacity while freeing manager capacity.

From the organization's perspective, delegation is a core component of operational resilience. A team that functions only when the manager is directly engaged is fragile — it cannot absorb the manager's absence, operate at speed, or scale. Organizations that invest in managers who delegate well build bench strength at every level because team members are continuously developing the skills and judgment that will make them future managers.

The AI dimension adds another layer of urgency. As artificial intelligence tools take over more of the routine, repeatable, and rule-based work that used to fill junior team members' calendars, managers must think carefully about how they structure the remaining work. High-value, judgment-intensive tasks — tasks that require human creativity, stakeholder relationships, and contextual reasoning — need to be distributed deliberately to develop the team's capability in these areas. Managers who do not delegate aggressively will find their teams underdeveloped precisely in the competencies that AI cannot automate. Delegation is not just a good leadership practice. In the AI era, it is a workforce strategy.

5.3 Why New Managers Struggle to Let Go

Understanding the psychological roots of under-delegation is the first step toward dismantling it. Most new managers were promoted because they were exceptionally good individual contributors — the best analyst, the strongest engineer, the most productive salesperson. That excellence was built on habits of personal execution: controlling quality, moving fast, knowing the details. Delegation requires temporarily giving up that control in service of a longer payoff that does not feel immediate.

The most common driver of under-delegation is identity. When a manager's sense of competence is still anchored in doing the technical work, stepping away from that work feels like becoming less valuable, not more. The answer to "What do I actually contribute?" feels uncomfortable if the honest answer is "I direct and develop other people." That discomfort leads managers to jump back in — not because the team cannot handle the work, but because handling the work is where the manager still feels capable and confident.

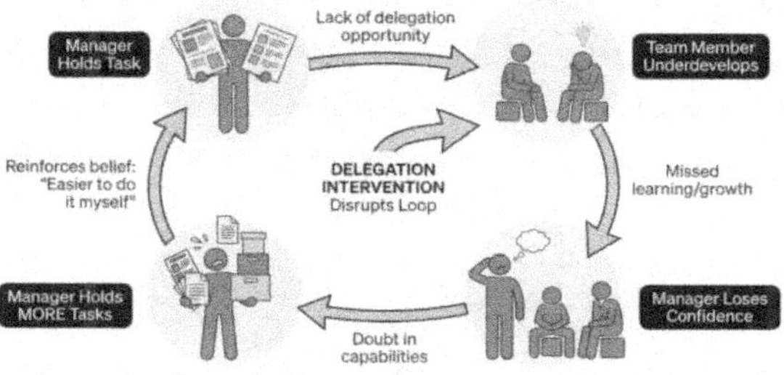

A second driver is the legitimate concern for quality. New managers worry — often correctly — that their team members will not do the work as well as they would. This concern is real, but it conflates two separate questions: whether the output will be perfect immediately and whether giving ownership is the right investment in the long term. If a manager only delegates work that is already guaranteed to succeed, no one learns anything. The cost of a slightly imperfect output today is often far lower than the cost of a team that has not developed its capabilities over a year.

Fear of being seen as lazy is a third, rarely acknowledged driver. Some managers believe that delegation means offloading work to appear unburdened, and they worry that their own manager will notice they are not producing outputs directly. This reflects a misunderstanding of what a manager's outputs actually are. A manager's outputs are the team's outputs. The manager who has delegated effectively and whose team is performing at a high level has produced more, not less, than the manager who is personally cranking out deliverables while their team sits underutilized.

Finally, the absence of a delegation system leads to poor decisions by default. When managers have no framework for deciding what to delegate, to whom, and how, they rely on intuition — and intuition tends to default toward retaining control. The answer is a repeatable framework that makes delegation decisions systematic rather than situational.

5.4 The Delegation Framework

Effective delegation is not a single decision but a structured process that moves through five stages: decide, select, brief, support, and review. Each stage requires specific manager behaviors and produces specific outcomes.

The first stage is deciding what to delegate. Not every task is a good candidate for delegation, but most tasks are better candidates than managers initially believe. A useful filter applies three tests: First, does this task require my unique authority, relationships, or decision-making level? If not, it is a candidate for delegation. Second, would completing this task develop a skill or build a team member's confidence? If so, it is an active reason to delegate, not just an absence of a reason to keep it. Third, is my time doing this task preventing me from doing something only I can do? If yes, the opportunity cost of not delegating is real. Apply these three tests systematically, and most of the work that piles up on a manager's plate turns out to be delegable.

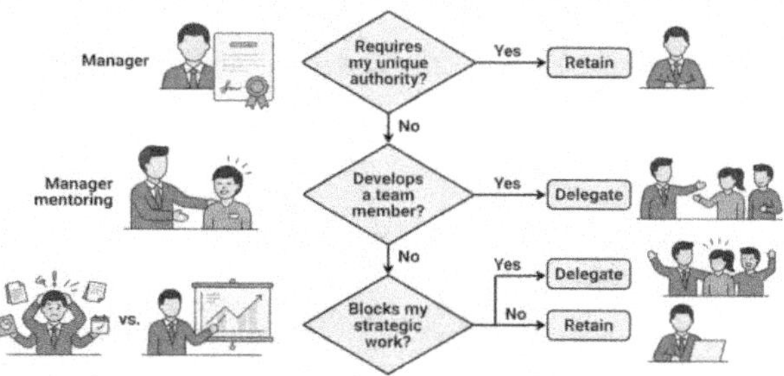

Systematically evaluate tasks using these three criteria to optimize executive time and empower team growth.

The second stage is selecting the right person. This is where the quality of delegation is either built or broken. The instinct is to give work to the person most capable of doing it perfectly. But perfect capability is not always the right criterion. Consider three dimensions: the team member's current skill level relative to the task's demands, their development goals, the growth opportunity the task represents, and their current workload and bandwidth. A task that is slightly above a team member's current level is an excellent development opportunity if the timeline allows for learning. A task given to an already-overloaded high performer is a setup for failure, not a reward.

The third stage is briefing with precision. The quality of a delegation brief determines the quality of the outcome more than any other single factor. A strong brief covers five elements: the purpose of the task and how it connects to broader objectives; the expected output and what success looks like; the constraints — timeline, budget, stakeholder sensitivities, non-negotiables; the level of authority the team member has — can they make decisions independently, should they consult before deciding, or should they only implement decisions you have made together; and the check-in cadence — when will you connect, and what does the team member need to report back versus handle autonomously.

Diagram 4.4: The Five-Element Delegation Brief

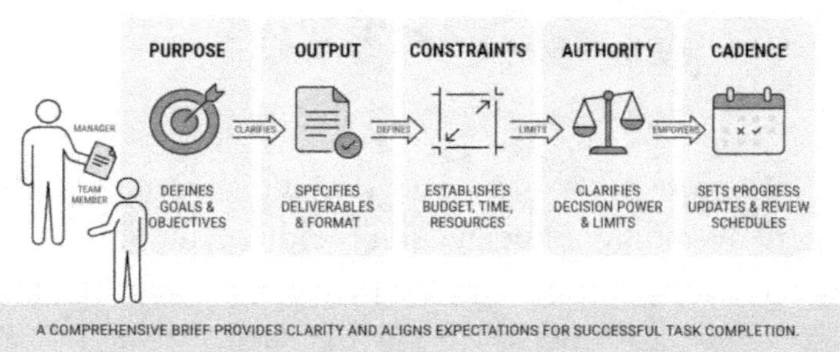

The fourth stage is support without interference. This is the hardest part for most managers, because the line between helpful guidance and undermining micromanagement is not always obvious. The distinguishing factor is whether the manager's involvement is responding to a team member's request for help, or whether the manager is inserting themselves based on their own anxiety about the outcome. If a team member asks a question, answer it. If a team member brings a problem, help them think through it. But do not check in more frequently than the agreed cadence, do not ask for updates that were not part of the brief, and do not revise outputs unless they are genuinely wrong — not just different from how you would have done it.

The fifth stage is review and recognition. When the work is done, deliberately close the loop. Give specific feedback on what went well and what could be stronger. If the output exceeded expectations, say so clearly and specifically. If there were gaps, address them as development opportunities rather than failures. The review stage is also where you gather information to improve your next delegation decision — what did this person learn, where did they get stuck, and what would make the brief more effective next time?

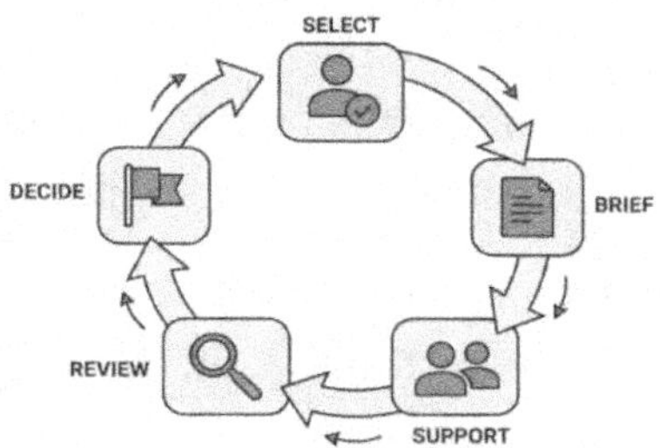

Effective delegation maximizes team productivity and fosters leadership development through a structured, repeating process.

5.5 Matching Tasks to Team Members

The skill challenge match is one of the most researched concepts in organizational psychology, and its core insight applies directly to delegation. When a task's demands are well above a person's current capability, they experience anxiety. When demands are well below their capability, they experience boredom. The most productive and developmental zone is when demands slightly exceed current capability — when the task requires the person to stretch but not break. Managers who calibrate delegation to this zone produce faster development and better outcomes than those who assign only tasks that people can do with certainty.

In practice, this means maintaining a working mental model of each team member's current capability across the dimensions relevant to their role. Technical skills are the easiest to assess because they are observable in outputs. Judgment, stakeholder management, and initiative are more nuanced — and these are precisely the dimensions where deliberate delegation can drive the most significant development. A team member who is technically strong but has never managed a stakeholder relationship benefits enormously from being delegated a task that requires it, with appropriate briefing and coaching support.

The growth dimension of delegation connects directly to each person's professional aspirations. When you know that a team member wants to move into a project management role, you can delegate tasks that give them relevant experience — leading a cross-functional meeting, managing a small budget, writing a project plan. This kind of intentional, development-oriented delegation is what separates managers who are mentors from managers who are task dispatchers. It signals to the team member that you understand their goals and are actively investing in their trajectory.

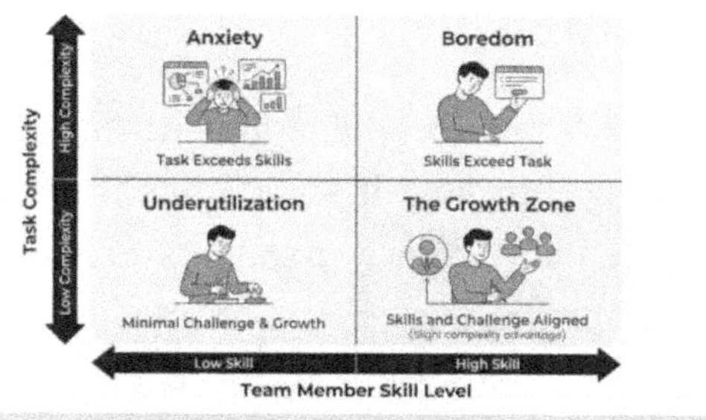

Workload distribution is a separate but related challenge. AI-powered project management tools increasingly make it possible to see across the team's commitments at a level of detail that would have required significant manual tracking in the past. Tools like Asana, Monday.com, and Microsoft Project now include workload heatmaps and capacity visualizations that, at a glance, show where each team member's time is committed and where there is room for additional ownership. Before making a delegation decision, consulting this data prevents the common mistake of assigning additional work to the team member who has done great work recently — who may already be at or near capacity — while overlooking a capable but less visible team member who has bandwidth and development potential.

5.6 Setting Expectations Without Micromanaging

The fear of micromanagement is so prevalent in management culture that it sometimes leads to its opposite: under-management, where managers are so worried about appearing controlling that they provide insufficient direction and then wonder why outcomes are inconsistent. The real goal is not to be hands-off by default. The real goal is to be calibrated — to match the level of involvement precisely to the team member's need for direction and the task's risk profile.

A useful framework for calibrating involvement is the four levels of autonomy. At level one, the team member gathers information and presents options, but does not act until the manager decides. This level is appropriate for high-stakes, low-experience situations — new team members working on critical tasks, or anyone entering genuinely unfamiliar territory. At level two, the team member develops a recommended action and waits for the manager's approval before proceeding. This level is appropriate for moderately experienced team members working on important tasks. At level three, the team member acts and then informs the manager. This is appropriate for experienced team members working within established boundaries. At level four, the team member acts completely independently, with no required reporting. This is appropriate for highly experienced team members working on familiar, low-stakes tasks.

The delegation brief should specify the level of autonomy that applies to each significant decision in the work. "For design decisions, you're at level three — do it and tell me. For budget changes above ten thousand dollars, you're at level two — recommend and wait for my approval." This precision prevents the frustration that arises when a team member makes a decision the manager expected to weigh in on, and it prevents the opposite frustration of waiting for approval on something the manager expected them to handle independently.

Progress visibility is not the same as control. A manager who wants to know where a project stands is not micromanaging — unless they respond to that information by interfering unnecessarily. Setting up a structured check-in — a weekly five-minute status update, a shared tracker that the team member updates — is not surveillance. It is operational clarity. The distinction is whether the check-in triggers manager action only when something is genuinely off track or triggers manager involvement by default. The former is professional oversight. The latter is micromanagement.

Diagram 4.7: The Four Levels of Autonomy

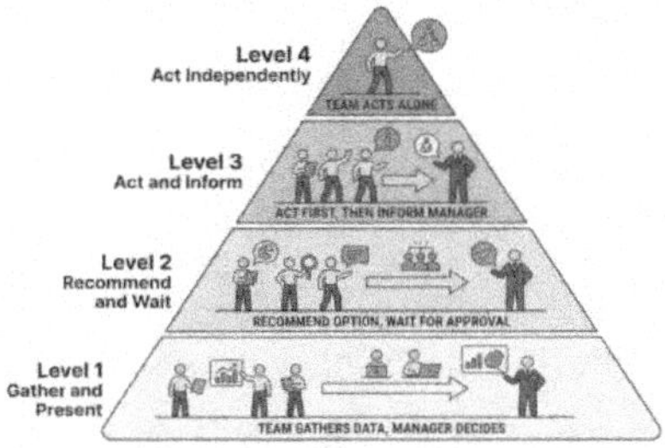

THIS PYRAMID ILLUSTRATES THE PROGRESSIVE INCREASE IN TEAM AUTONOMY AND DECISION-MAKING POWER AS MANAGER INVOLVEMENT DECREASES.

5.7 AI Tools for Smarter Delegation

The generation of AI-enhanced project management and workflow tools that has emerged over the last several years has made delegation decisions more data-informed than ever. Rather than relying entirely on memory and informal observation to know who has capacity and who is skilled at what, managers can now draw on structured data to make better assignment decisions.

Workload visibility is the most immediate application. Tools like Asana Intelligence, Monday.com's AI-powered workload view, and Jira's capacity planning features aggregate time commitments across the team and surface patterns that would otherwise be invisible. A manager looking at a capacity heatmap can see that two team members are consistently running above 100% of planned capacity. In contrast, two others are consistently below 80% — information that should directly shape delegation decisions over the coming weeks.

Skills mapping is a higher-order application. Platforms like Workday, Eightfold AI, and SAP SuccessFactors now maintain skills profiles for every team member, updated through a combination of self-reporting, manager assessment, and analysis of work history. When a manager needs to assign a task that requires specific expertise — stakeholder communication, SQL querying, regulatory compliance knowledge — a skills search surfaces the team members most suited to that task and, equally usefully, those for whom it would be the right development stretch.

Task automation fundamentally changes the calculus of delegation. When AI tools can draft routine communications, generate status summaries, analyze datasets, and compile reports, the question of what to delegate changes. The lowest-value work that used to occupy junior team members' time is increasingly being automated. This is, broadly, good news — it means team members' time is freed for more judgment-intensive work. But it creates an obligation for managers to think deliberately about what replaces that routine work in team members' development portfolios. If AI is writing first drafts, junior team members need explicit opportunities to develop editing judgment, conceptual thinking, and stakeholder communication — skills that are not developed by passively reviewing AI output, but by owning meaningful decisions.

Prompt engineering for delegation support is an underused application of generative AI. A manager preparing a delegation brief for a complex project can use an AI assistant to stress test the brief — asking it to identify ambiguities, surface potential failure modes, or suggest clarifying questions the team member might need answered. The AI does not write the delegation brief, but it can significantly strengthen it by challenging the manager's assumptions before the brief is delivered.

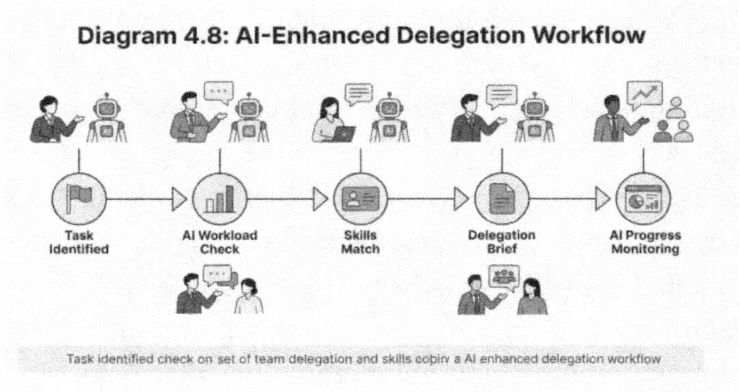

5.8 Building a Culture of Ownership and Autonomy

Delegation done well, consistently, and over time, produces something more valuable than efficiently distributed work — it produces a team culture characterized by ownership mentality, autonomous problem-solving, and mutual accountability. These cultural qualities do not emerge from any single delegation decision; they are built through the accumulation of experiences where team members have been trusted with meaningful work, supported appropriately, and recognized authentically for their results.

Ownership mentality is the internalized belief that the work you have been given is genuinely yours to drive, not yours to do until the manager takes it back, not yours to execute under constant supervision. Still, yours to own from start to finish, including decisions, setbacks, and adjustments. Building this belief requires consistency. If a manager delegates a project and then reasserts control at the first sign of difficulty, the team learns that delegation is conditional — that ownership is revoked when things get hard. This teaches learned helplessness faster than any other management behavior. Managers must be willing to let team members work through difficulty rather than reflexively rescuing the situation.

Autonomy at the team level requires that each person understands not just what they are supposed to do, but why it matters and what success looks like — so that they can make good decisions in situations the delegation brief did not specifically anticipate. This is the difference between completing a mechanical task and genuine ownership. A team member who understands the goal well enough to navigate unexpected obstacles is operating autonomously. A team member who stops moving the moment they encounter something not covered in the brief is not yet there. The prescription is a more thorough briefing, not less delegation.

Mutual accountability — where team members hold each other to commitments, not just the manager holding them — is the highest expression of a healthy delegation culture. It emerges when team members understand how their work connects to others', when they feel responsible to their peers as well as their manager, and when the team's shared goals are clear enough that falling short of them is meaningful to everyone. AI tools that provide shared visibility into project status and progress contribute to this dynamic by making the interdependencies among team members' work concrete rather than abstract.

A manager who wants to accelerate the development of an ownership culture should create deliberate opportunities for team members to experience the full arc of ownership — including the uncomfortable parts. That means assigning stretch projects with real stakes. It means being available for coaching without doing the work. It means accepting outputs that are different from what you would have produced but are still successful. And it means celebrating the ownership itself, not just the outcome — recognizing the team member who navigated a difficult situation independently as much as the one who produced a flawless deliverable with close guidance.

5.9 The Manager's Self-Assessment: Are You Under-Delegating?

Self-awareness is a precondition for improving delegation. Before a manager can change their delegation habits, they need an honest accounting of where those habits currently stand. Most managers who under-delegate believe, at some level, that they are appropriately involved — that the work they retain genuinely requires their involvement, and that the standards they enforce through direct engagement are necessary for the team's success. This belief is sometimes correct and far more often a comfortable rationalization.

A useful self-assessment starts with a time audit. For one week, document every task you work on and categorize it according to a simple framework: tasks that only you can do given your role, authority, and relationships; tasks that a team member could do with appropriate briefing and support; and tasks that a team member should have done independently but that you took over. The third category is the most diagnostic. If you find yourself regularly completing tasks in that third category, you are under-delegating — and the pattern will not change until you address it explicitly.

A second diagnostic is the team's question pattern. When team members bring you questions, what kinds of questions are they? Questions about strategic direction, organizational politics, and stakeholder relationships are the manager's domain — answering these is part of the job. Questions about how to format a report, which tool to use for a particular analysis, or whether a draft communication is ready to send are indicators that the team member has not been given sufficient ownership of that work, or that previous experiences have taught them that checking in is safer than making decisions independently. The proportion of these questions in a typical week is a direct indicator of the team's delegation health.

A third diagnostic is what happens when you are out. Take note of what requires your personal handling when you are on vacation or in back-to-back meetings that make you unreachable for a day. The list of items that stall, that team members hold for your return, and that produce a queue of unresolved issues is a direct inventory of work that has not been genuinely delegated. Work that has been truly delegated — with authority, context, and accountability — continues moving in the manager's absence.

The goal of this self-assessment is not guilt but clarity. Every manager under-delegates in some areas, and the areas where you are holding on too tightly are often the areas that matter most to your professional identity. Seeing that pattern clearly, without judgment, is the first step toward changing it. The next step is choosing one item from your self-assessment — one task or decision you have been retaining — and delegating it this week with a complete brief and a genuine transfer of authority.

5.10   Delegation as a Development Strategy

The most sophisticated use of delegation is not workload management — it is talent development. A manager who thinks about delegation through the lens of development designs their assignment patterns intentionally over time, treating the team's project portfolio as a curriculum that systematically builds the capabilities each person needs to reach their next level of performance and their next stage of career progression.

Development-oriented delegation requires understanding two things about each team member: where they are now, in terms of the skills and experiences they have already demonstrated, and where they want to go, in terms of the professional goals and aspirations they have expressed. With both of those data points in hand, a manager can identify the gap and select delegation assignments that bridge it. A team member who is technically strong but has never had budget responsibility benefits from being given a small project budget to manage. A team member who excels in individual work but wants to develop leadership skills benefits from being given a task that requires coordinating the work of one or two peers. A team member who has expressed interest in moving into client management benefits from being brought along to stakeholder conversations and eventually given ownership of a client relationship.

This kind of intentional development sequencing has a compounding effect. Over twelve months of deliberate assignment decisions, a manager can produce meaningful capability growth in every member of the team — not the performance management kind of growth that addresses deficits, but the development kind of growth that expands range and prepares people for greater scope. The team that emerges from that year is materially more capable than the one that entered it, and the individual team members are more engaged because they can see their own trajectory.

Development-oriented delegation also has retention implications that are difficult to overstate. When team members experience that their manager is actively investing in their development through the work they are given — not just telling them in one-on-one conversations that their growth matters, but demonstrating it through actual assignment decisions — they develop a loyalty to the manager and the team that is hard to replicate with compensation alone. The manager known for developing their team attracts talented people who want to grow, retains them during periods when external compensation is more attractive, and builds a reputation that makes every subsequent recruiting conversation easier.

Tracking development progress against a deliberate plan transforms delegation from an art form into a managed practice. A simple document for each team member —

listing their current development priorities, the assignments being used to address them, and the progress being made — creates accountability on both sides and gives the manager the information they need to ensure that development opportunities are being distributed equitably across the team. Without tracking, the most vocal and confident team members tend to receive the most visible development assignments. In contrast, quieter team members who may have equal or greater potential receive fewer stretch opportunities by default. Intentional tracking counteracts this pattern.

5.11   Delegation Across Experience Levels

Delegation is not a uniform practice — it requires meaningful adaptation based on the experience level and confidence of the team member receiving the work. A new graduate in their first professional role and a senior team member with a decade of relevant experience both deserve well-structured delegation. Still, the nature of that structure looks fundamentally different. Applying the same approach to both is one of the most common mistakes managers make when they first develop a delegation practice.

For team members who are early in their careers or new to a type of work, the briefing stage needs to be substantially more detailed, and the check-in cadence substantially more frequent. This is not micromanagement — it is appropriate scaffolding. When someone has not yet developed the pattern recognition that comes from years of experience, they genuinely need more context, more explicit criteria for what success looks like, and more frequent opportunities to check their understanding against the manager's expectations before the work is too far along to correct. A brief that would be entirely sufficient for a ten-year veteran — "draft the executive summary, you know what we need" — leaves a first-year team member without the information they need to produce anything close to the intended output.

The appropriate response to this is not to assign less important work to early-career team members. It is to invest more in briefing and support structures so they can successfully own meaningful work earlier than they otherwise would. The manager who gives a junior team member a real challenge with a comprehensive briefing, calibrated check-ins, and genuine coaching support through the difficulty develops that person significantly faster than the manager who waits until the junior team member is "ready" by some standard of independence that can only be built through ownership experience. Readiness and ownership are not sequential — they develop simultaneously.

For highly experienced team members, the most important adaptation is to step back further and resist the impulse to add guidance that the person does not need. Experienced team members often report feeling insulted by over-briefing — by being told things they already know, as if they were not trusted to know them. This is a form of disrespect, even when it is unintentional. The delegation brief for an experienced team member should primarily cover context and authority level, not how to do the work. The manager's job in supporting an experienced team member is to remove obstacles, provide organizational air cover, and stay out of the way of someone more capable than the manager in the specific domain of the delegated work.

Managing a team spanning multiple experience levels requires differentiated judgment in every delegation decision. The same project may require dense briefing and close support for one team member and a single-sentence summary for another. This differentiation is not unfair — it is responsive to actual need. What would be unfair is treating everyone the same when their situations differ, which, in practice, means either leaving junior team members unsupported or condescending to experienced ones. The manager who calibrates their delegation approach to the individual, rather than applying a single template to everyone, produces better outcomes for every member of the team and builds the kind of relationship where team members feel genuinely seen and understood.

5.12   Manager's Delegation Checklist

Before delegating any significant task, confirm you have addressed each of the following.

Purpose and context provided. The team member understands why this work matters, how it connects to team goals, and what happens downstream of their output.

Output is defined specifically. You have described what a successful deliverable looks like in concrete terms — not "a good report" but "a one-page executive summary covering the three metrics the director asked about, written for a non-technical audience, ready by Thursday EOD."

Authority level specified. The team member knows explicitly which decisions they own, which they should bring to you, and which require higher-level approval. Do not leave this to assumption.

Constraints surfaced. Budget limits, stakeholder sensitivities, format requirements, and non-negotiable quality standards have been communicated before the work begins, not discovered after the first draft.

Check-in cadence established. You have agreed on when and how often you will connect — and you have committed to using those check-ins, not adding unscheduled touchpoints based on your own anxiety.

Skills match confirmed. You have verified that the task is a reasonable match for the team member's current capabilities — challenging enough to be developmental, not so difficult as to set them up to fail without support.

Workload capacity checked. Before assigning the task, you have confirmed that the team member has the capacity to take it on without sacrificing quality on existing commitments.

Recognition plan in place. You have thought about how you will recognize this team member's ownership of the work — both during the process and when it concludes — not just whether the output was delivered.

Development feedback planned. You have prepared to give specific, actionable feedback after the work is done, regardless of outcome, that helps this person do the next thing even better.

No rescue reflex. You have committed to allowing the team member to work through difficulties rather than intervening at the first sign of struggle, and you know how to distinguish situations that require your intervention from those that require your restraint.

5.13   The Multiplier Manager

Delegation is not about getting work off your plate. It is about building a team that can carry more weight, move faster, and develop deeper capability than any single manager working alone could achieve. Every task you delegate intentionally — with a precise brief, the right level of authority, and genuine follow-through on the review — is an investment that compounds. The team member gets stronger. The team's collective capacity expands. And you free the time and mental bandwidth to do the strategic, relational, and organizational work that only a manager can do.

The managers who struggle most with delegation are almost always the ones whose identities are still anchored in individual performance. The shift — from "I am productive when I am producing outputs" to "I am productive when my team is producing outputs because of how I have led them" — is one of the most important cognitive transitions in a management career. It does not happen all at once. It happens through the accumulation of small delegation decisions, each one slightly more trusting than the last, until the new identity takes root.

AI tools accelerate this transition by providing the visibility and support structures that make delegation feel less like releasing control and more like maintaining intelligent oversight. When you can see the team's workload in a dashboard, track project progress in a shared system, and receive AI-generated summaries of where each initiative stands, delegation becomes less of a leap of faith and more of a managed transfer of ownership. The tools do not replace the trust — nothing does —, but they make it easier to extend.

Start this week by taking on one task you have been holding for someone on your team. Brief them thoroughly. Specify their level of authority. Set a check-in for Thursday. And resist the urge to intervene before Thursday arrives. That one decision, replicated consistently over months, is how managers become multipliers.

6 Setting Goals and Driving Performance

6.1 Scenario

The end-of-year performance cycle arrives like a recurring natural disaster at many organizations — dreaded by managers, resented by team members, and largely disconnected from the actual work that happened during the year. Marcus, a second-year manager leading a six-person marketing analytics team, knows this feeling intimately. He spent the better part of three days in December drafting performance reviews, pulling from memory and a handful of archived emails, trying to reconstruct a year of work into structured summaries. Half his team had goals written in January and never revisited. Two team members took on substantial new responsibilities in September when a colleague departed, but their official goals reflected none of that work.

When he sat down to deliver the reviews, the conversations felt retroactive and slightly arbitrary. Two people were surprised by the developmental feedback — not because they disagreed with it, but because they were hearing it for the first time rather than recognizing it as a consistent theme from ongoing conversations. One high performer left the conversation less motivated than she had arrived, frustrated that her expanded scope over the final quarter had not been reflected in her rating. Marcus worked hard on those reviews, but the process was doing more harm than good because the foundation — clear, current, shared goals connected to ongoing feedback — had never been built.

This chapter is about building that foundation. Goal-setting and performance management are not annual paperwork exercises. They are continuous leadership practices that, when done well, align team energy with organizational priorities, surface obstacles before they become failures, and give every person on the team a clear understanding of what excellent work looks like and how close they are to it. AI tools are beginning to transform this practice at every stage — from goal alignment to progress tracking to feedback analysis — but the human judgment at the center remains the manager's responsibility and the manager's opportunity.

6.2 Why Goal-Setting Is the Manager's Anchor

Clear goals are the single most powerful predictor of team performance, and they are the manager's primary responsibility to establish. When team members do not know what success looks like, their effort becomes fragmented. Each person optimizes for what feels important to them individually, which may or may not align with what the team or organization actually needs. The result is busyness without direction, activity without impact, and the particular frustration of a team that is working hard but not moving toward anything coherent.

The research on goal-setting is unusually consistent. Specific, challenging goals produce higher performance than vague or easy goals. Goals that are written down and tracked produce better results than goals that exist only in memory. Goals that team members have helped set produce higher commitment than goals handed down without discussion. Goals connected to a visible organizational purpose generate higher intrinsic motivation than those that feel arbitrary or purely managerial. None of this requires management theory expertise to implement — it requires attention and consistency.

Diagram 5.1: The Goal-Setting Impact Chain

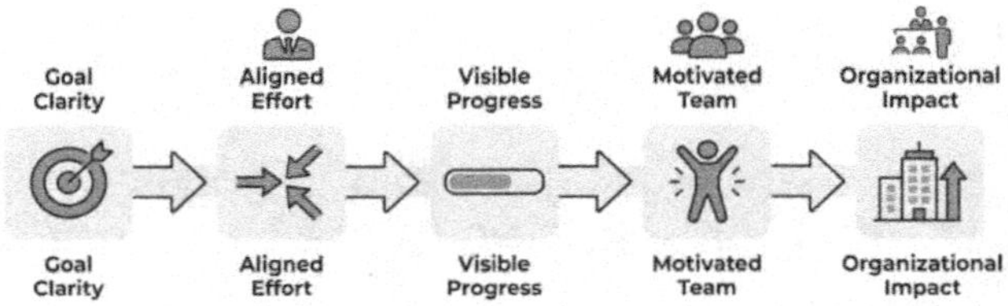

The Goal-Setting Impact Chain e a complete to sn varged effortor motivated team summing up an executive business tone.

The manager's role in goal-setting is more active than most goal-setting advice acknowledges. It is not sufficient to hold a goal-setting meeting in January and then wait for the end of the year to evaluate results. Goals set without regular review become outdated within months as priorities shift, business conditions change, and individual circumstances evolve. A goal that was ambitious in January may be either irrelevant or insufficiently challenging by May. The manager who updates goals in response to reality — not to lower standards, but to keep them meaningful and current —

maintains a team that is always working toward something real.

Goal-setting also communicates what the manager values, which is itself a powerful signal. When a manager sets goals that are purely output-oriented — number of deals closed, volume of code shipped, number of tickets resolved — the team receives a clear message that output is what matters and everything else is subordinate. When goals encompass quality, development, collaboration, and process improvement, the team receives a more complete and accurate picture of what excellent performance looks like. The goals themselves are an expression of the manager's theory of what great work means on this team.

6.3 Goal-Setting Frameworks

Two frameworks dominate practical goal-setting in modern organizations: OKRs (Objectives and Key Results) and SMART goals. Each has genuine strengths, and the choice between them — or the decision to use both in different contexts — depends on the team's operational rhythm, the organization's culture, and the nature of the work.

OKRs were popularized by Intel and later by Google and have since become the standard goal-setting framework in technology companies and growth stage organizations. The structure is deceptively simple: an Objective states a qualitative, inspiring direction — where you want to go. Key Results are the quantitative metrics that indicate whether you have arrived. A strong Objective connects to the mission and generates energy: "Build a data analytics capability that drives commercial decisions" is a real Objective. "Improve our analytics" is not. Key Results must be specific, measurable, and time-bound: "Reduce average dashboard load time from fourteen seconds to under three seconds by Q2" is a strong Key Result. "Improve dashboard performance" is not.

Diagram 5.2: OKR Architecture

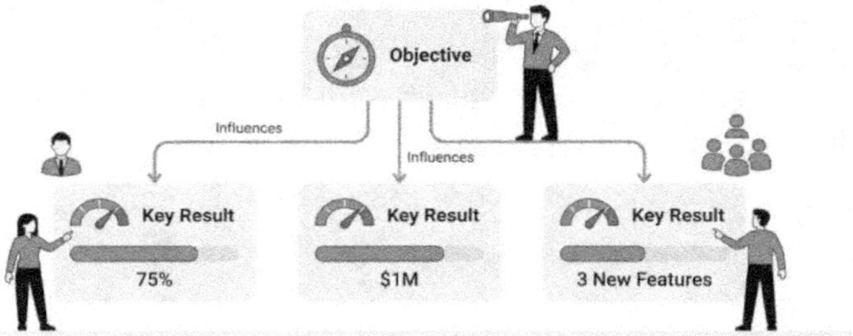

A clear architecture aligning high-level objectives with measurable key results for strategic goal attainment.

The power of OKRs comes from their nested nature. Organizational OKRs cascade to team OKRs, which cascade to individual OKRs — creating an explicit visual and logical connection between what each person is working on and what the organization is trying to accomplish. This alignment is not just administratively useful; it is motivationally powerful. Team members who can see how their Key Results contribute to a team Objective that contributes to an organizational Objective work with a sense of purpose that is difficult to manufacture by other means.

SMART goals — Specific, Measurable, Achievable, Relevant, and Time-bound — are older and more widely used across organizations that do not operate in the OKR framework. Their strength is their practicality: the SMART filter is easy to apply to any task or project, and it prevents the vagueness that makes goals unactionable. A goal that passes the SMART test is ready to be tracked and evaluated. A goal that fails any of the five tests needs more work before it becomes a genuine commitment.

The most effective approach for most managers is to use OKRs for team-level strategic goals that cascade from organizational priorities, and SMART goals for individual project-level commitments that may be more tactical. This gives the team a strategic horizon — the Objectives and Key Results that define where the team is going over the quarter or year — alongside the clear, practical milestones that govern day-to-day work. The two frameworks are complementary rather than competing.

Beyond OKRs and SMART goals, a third concept worth incorporating is the distinction between outcome goals and process goals. An outcome goal specifies a result: "Increase customer satisfaction score from 78 to 85 by the end of Q3." A process goal specifies a behavior or practice that is expected to produce the result: "Conduct a customer call within 24 hours of each support ticket closure." Outcome goals clarify the target; process goals specify the actions the manager believes will reach it. Holding both kinds of goals simultaneously keeps the team focused on what they can control — their behaviors and practices — while maintaining clarity about what success ultimately looks like.

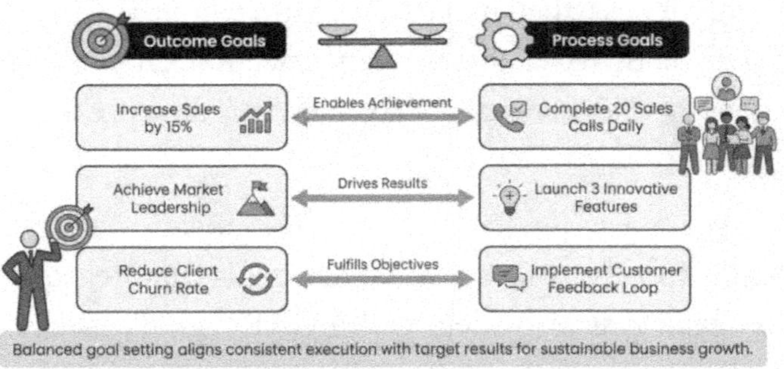

6.4 Aligning Team Goals with Organizational Strategy

The value of a goal is directly proportional to the clarity of its connection to something the organization genuinely cares about. Goals that exist in isolation — disconnected from organizational priorities, invisible to stakeholders outside the team, and unconnected to resource allocation decisions — are goals in name only. They create local activity without contributing to organizational momentum. One of the manager's most important functions is to be the translator between organizational strategy and team-level goals.

Translation in this context requires two things: understanding the organizational strategy well enough to know what it actually asks of your team, and understanding your team's work well enough to identify where it can contribute most directly. This understanding cannot be delegated. It requires the manager to engage actively with organizational communications — earnings calls, if you are in a public company context; all-hands presentations; and strategic planning documents — and to synthesize that information into implications for the team's priorities and goals.

When goal-setting season arrives, present the connection explicitly. Show the team, in concrete, specific terms, how the goals you are proposing for the quarter connect to the organizational priorities communicated from the top. "The company has committed to reducing customer churn by fifteen percent this year. Our team's contribution to that commitment is to improve the reliability of our analytics dashboards, because sales leaders have told us that unreliable reporting is among the top five reasons customers escalate. So our Key Result — 99.9% dashboard uptime — is not an internal engineering metric. It is a commercial commitment." This kind of explicit connection is energizing in a way that abstract goal alignment language is not.

Diagram 5.4: The Strategic Alignment Cascade

Ensuring daily actions contribute directly to achieving top-level corporate objectives.

Mid-cycle goal alignment is as important as initial alignment. When the organization changes its priorities — as organizations frequently do — the manager must revisit team goals proactively and adjust. Continuing to measure a team against goals that are no longer aligned with organizational priorities wastes the team's energy on work that has lost its value and creates a confusing, demoralizing situation in which a team has performed well against their stated goals but produced little of what the organization actually needed. The manager who recalibrates goals in real time keeps the team's operations relevant.

6.5 The Performance Management Cycle

Performance management is not a once a year event. It is a continuous cycle comprising four interconnected phases: planning, monitoring, reviewing, and developing. Each phase depends on the quality of the previous one, and each contributes to the next. Managers who skip phases or compress this cycle into an annual review feel the consequences in the quality of their feedback, the predictability of their outcomes, and the engagement of their team.

The planning phase — typically at the start of a performance period — is where goals are set, expectations are documented, and the basis for evaluation is established. This is the most important phase and the most frequently underinvested. A planning conversation should cover not just what goals are being set, but why those specific goals were chosen, what success looks like at each level of the rating scale, and what support the manager will provide to help the team member reach their goals. A planning conversation that ends without the team member clearly understanding how they will be evaluated is not finished.

The monitoring phase is ongoing. It encompasses every one-on-one conversation, every check-in on a project, every informal observation of how work is progressing. Good monitoring is not surveillance — it is the manager paying enough attention to know when a team member is on track, when they are struggling, and when they need developmental input. Monitoring also includes tracking the formal metrics established in the planning phase, updating them regularly, and making that data visible to the team member so they are never surprised by their progress status.

The reviewing phase is the formal periodic evaluation — typically quarterly or annually. The preceding monitoring phase almost entirely determines the quality of this phase. If the manager has been having honest, substantive conversations throughout the period, the review conversation is a summary and synthesis of things the team member already knows. If the manager has been avoiding difficult observations during the year, the review becomes the uncomfortable delivery of news that should have been communicated months earlier. This is why surprise in a performance review is a management failure, not a performance management mechanism.

The Performance Management cycle is an ongoing process designed to align employee and organizational goals.

The developing phase is where the review feeds forward into growth. After evaluating performance, the manager and team member should immediately shift to forward-looking questions—what skills should be developed, what opportunities are available, and what support the manager will provide in the next period. Development planning without this forward-looking conversation produces accurate assessments but does not drive growth. The cycle's value is in the feedback loop — each period's learning informs the next period's planning.

6.6 Performance Reviews That Motivate

A performance review conversation that leaves a team member feeling demoralized has failed, regardless of how accurate its content was. The goal of a review is not to deliver a verdict — it is to create shared understanding that drives future performance. The distinction matters enormously for how the conversation is structured, what language is used, and how the manager prepares.

Preparation is the non-negotiable prerequisite. A manager who enters a review conversation without having reviewed the year's data, refreshed their memory on specific examples, and prepared their key messages is not prepared to give a review — they are prepared to improvise, and improvisation in this context produces vague, unmemorable, and often inconsistent feedback. Take the time to gather concrete examples from across the performance period, consider the team member's perspective on their own performance, and think through the developmental message you want to leave them with.

The review conversation should begin with the team member's self-assessment, followed by the manager's. Asking "How do you think this period went?" before delivering your evaluation accomplishes several things simultaneously: it surfaces any significant misalignments in perception that need to be addressed head-on rather than overrun; it gives the team member the experience of being heard rather than evaluated; and it often provides useful context that changes how the manager frames their feedback. A team member who immediately acknowledges a known performance gap before you raise it is already in a growth mindset — that conversation should proceed differently than one where the same team member is surprised by the same observation.

Specific, behavioral, and impact-focused feedback is more motivating than summary ratings. "You consistently delivered the weekly analytics reports ahead of schedule, which meant the sales team always had current data for their Monday briefings — that directly contributed to three of our top five client renewals this quarter," says a team member, something meaningful and actionable. "You did a good job with reporting" tells them almost nothing and leaves them no better equipped to understand what to sustain in the future. The same principle applies to developmental feedback: "I noticed that when you present to the executive team, you tend to lead with methodology rather than conclusions, and I have seen a few stakeholders disengage as a result — I want to work with you on shifting to a headline first structure" is actionable. "You need to improve your executive communication" is not.

Diagram 5.6: The SBI Feedback Structure

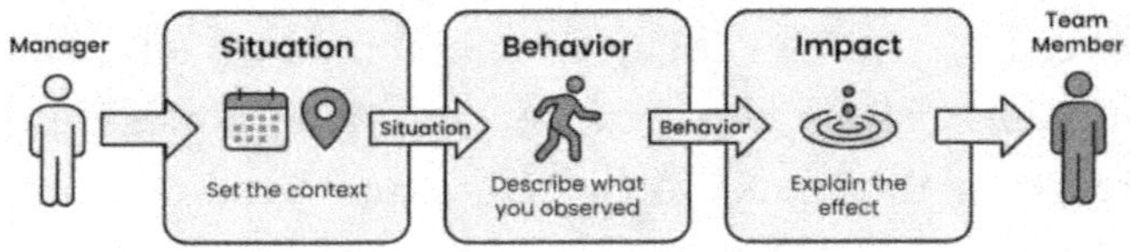

Utilize the Situation-Behavior-Impact model for effective leadership feedback sessions.

Ending the review with a forward-looking plan — not just a rating and a signature — transforms the experience from judgment to development. "Here are the two things I most want to see you grow in next period, and here is what I am going to do to support that growth," closes the review with energy and direction. The team member leaves with a clear understanding of what success looks like in the next period, which is precisely the foundation the planning phase needs.

6.7 Addressing Underperformance with Confidence

Underperformance conversations are among the hardest things managers are asked to do, and avoidance is the most common failure mode. Managers delay, soften their language to the point of meaninglessness, hint around the problem without naming it directly, and then discover at year-end that the team member had no idea there was a serious concern about their performance. This pattern is not kindness — it is a disservice to the team member, who deserved the honest feedback that would have given them a real chance to improve.

The first rule of addressing underperformance is to be direct without being harsh. A conversation that opens with "I need to talk with you about something serious, and I want to be direct with you because I think you deserve clear feedback" is both kinder and more effective than one that buries the concern in qualifications and softening language. Team members who receive vague feedback cannot improve because they do not know specifically what needs to change. Team members who receive direct, specific behavioral feedback have a genuine path forward.

The Performance Improvement Plan — often called a PIP — is a formal mechanism that documents specific performance gaps, clear expectations for improvement, a defined timeline, and the support the organization will provide. PIPs are often treated as the beginning of a termination process, and in some organizational cultures, that is, in fact, what they have become. But a well-constructed PIP, delivered with genuine investment in the team member's success, is a structured opportunity for recovery. The manager's posture in delivering a PIP matters as much as its contents: "I have put this plan together because I want you to succeed in this role, and I need you to understand exactly what success looks like so you have a real chance to achieve it" is a different message than one that feels like paperwork for a foregone conclusion.

A manager who addresses underperformance early — at the first sign of a pattern, not after months of frustration — gives the team member more time to recover, gives the team more time without the drag of unresolved performance issues, and gives themselves the relief of having done their job. Delayed difficult conversations do not get easier with time. They accumulate damage while the manager waits for them to become unavoidable.

Throughout any underperformance process, documentation is the manager's professional protection and the team member's fair record. Document the specific behaviors observed, the feedback given, the support offered, and the team member's responses. This documentation serves multiple purposes: it ensures that both parties have a consistent record of what was discussed and provides legal and HR protection if a formal process becomes necessary. It holds the manager accountable for the commitments they made to support improvement.

6.8 AI-Powered Performance Analytics

The arrival of AI-enhanced HR and people analytics platforms has given managers access to a level of performance data visibility that was previously available only to large organizations with dedicated analytics teams. These tools do not replace the human judgment at the center of performance management — but they sharpen it considerably by surfacing patterns, identifying trends, and reducing the cognitive load of tracking multiple team members' progress simultaneously.

At the most basic level, modern HR platforms like Workday, BambooHR, and Lattice provide structured goal tracking, making the planning through review cycle substantially more manageable. Goals set in the planning phase are documented in the system, progress can be updated continuously, and the review conversation draws from a shared, current record rather than from reconstructed memory. The administrative overhead of performance documentation drops, and the quality of the data available at review time rises.

Productivity analytics tools like Microsoft Viva Insights and similar platforms aggregate collaboration and work pattern data — meeting loads, after hours work, communication volume, collaboration network breadth — to surface potential wellbeing and engagement risks before they manifest as turnover or performance problems. A manager who sees that a team member's after hours activity has spiked significantly over several weeks has an early indicator worth exploring in a one-on-one conversation: not as surveillance data, but as a signal that warrants a well-being check.

Sentiment analysis applied to employee survey data is another emerging application. Platforms like Glint, Culture Amp, and Qualtrics EmployeeXM use natural language processing to analyze open text survey responses and identify themes and sentiment patterns across the team. A manager who sees a recurring theme in their team's survey comments about clarity of priorities — even if the comments are worded differently by different team members — has actionable information on where their communication and goal-setting practices might be improved.

Predictive attrition tools, now available on enterprise HR platforms, use machine learning to analyze indicators such as engagement scores, compensation benchmarking, tenure patterns, and career progression trajectories to flag team members who are statistically more likely to leave. This information is not deterministic — it identifies risk, not certainty — but it allows managers to have proactive retention conversations rather than reactive exit interviews. A manager who knows that a high performer is showing attrition risk indicators can explore their development aspirations, compensation concerns, and engagement drivers before those concerns become a resignation.

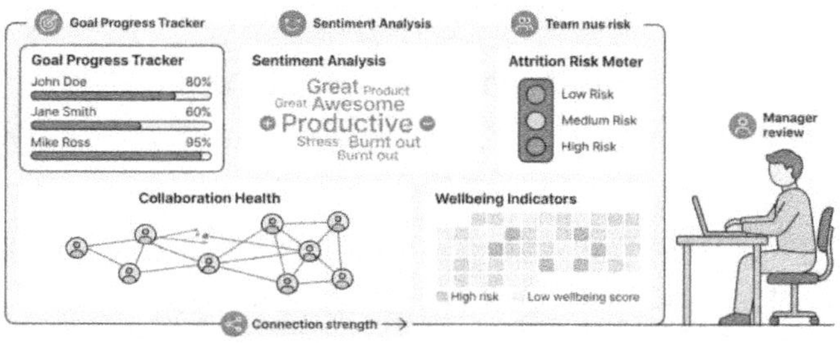

AI provides actionable insights into team productivity, engagement, and overall performance metrics.

The ethical dimension of performance analytics deserves explicit attention. Data collected about team members carries responsibilities. Before deploying any people analytics tool, managers should ensure that team members understand what is being tracked and how it will be used; that the data is being used to support development and wellbeing, not to build surveillance files; and that any inferences drawn from the data are shared with the team member as part of a two-way conversation, not treated as conclusions reached without their input. AI-powered analytics should inform the manager's judgment — they should not replace the human conversation that remains at the center of effective performance management.

6.9 Using Data to Celebrate Wins and Identify Trends

Performance data is most commonly discussed to identify problems. But the same data, used proactively, is equally powerful as a tool for recognizing achievement, reinforcing positive patterns, and building the motivation that sustains high performance over time.

Specific, data-anchored recognition is more powerful than generic praise. When a manager can say, "Your work on the Q3 client analysis contributed directly to three renewal decisions that represented roughly eight hundred thousand dollars in retained revenue — I looked at the timeline, and your analysis was the turning point in each of those conversations," the recognition is concrete, credible, and memorable. The team member knows exactly what behavior to repeat. When the same recognition is framed as "Great work this quarter, really strong contribution," it is pleasant but not actionable — the team member has no clear signal about what specifically was valuable.

Diagram 5.8: Recognition That Compounds

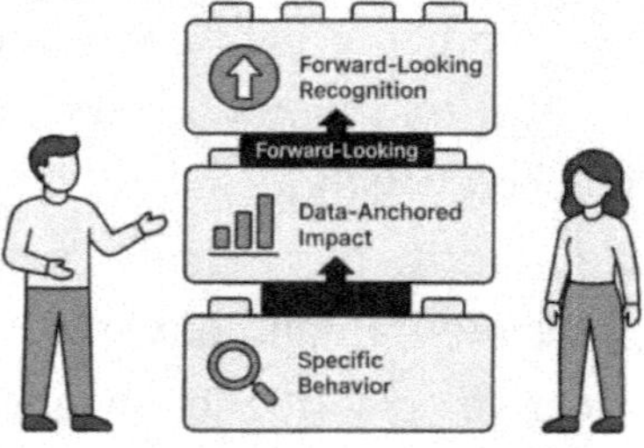

Recognizing specific behaviors and their data-anchored impact leads to forward-looking, impactful employee recognition.

Trend identification is a forward-looking use of performance data. A manager who notices that a team member's quality metrics have been declining gradually over three months has information worth acting on now—not after the decline becomes a crisis. The conversation is completely different at month three than at month nine: at month three, it is a coaching conversation about what might be happening and how the manager can support a course correction. At month nine, it is a performance management conversation with a much longer tail. Early trend identification, enabled by consistent data tracking, keeps more conversations in the development zone and fewer in the performance management zone.

Team-level trend analysis extends this capability to the team. A manager who tracks the team's aggregate output metrics over time can identify whether the team is on a sustainable trajectory, whether particular types of work are consistently overrunning estimates, whether team output drops during certain periods that might correspond to specific challenges, and whether the team's capacity has grown or contracted relative to its goals. This analysis is now substantially easier with AI tools that aggregate and visualize team metrics without requiring the manager to compile spreadsheets manually. The operational clarity that results from this visibility is not just useful for performance management — it is essential input for capacity planning, hiring decisions, and goal-setting.

6.10   The One-on-One as a Performance Management Tool

The weekly or biweekly one-on-one meeting is the most underutilized performance management tool available to managers. Most managers treat the one-on-one as a status update — a recurring calendar slot where the team member reports what they are working on and the manager nods and offers occasional input. This is a significant missed opportunity. A well-structured one-on-one is the primary mechanism through which a manager monitors progress in real time and delivers developmental feedback. At the same time, it is still relevant, identifies obstacles before they become crises, and builds the relational foundation that makes all other performance management conversations more effective.

The structure that makes one-on-ones useful is simple but requires consistency. Roughly one-third of the time is for the manager: updates on organizational developments, context on changing priorities, and any feedback or recognition the manager has been accumulating since the last meeting. Roughly two-thirds of the time belongs to the team member: their current challenges, the questions they need answered, the developmental topics they want to explore, and the progress they want to report. This ratio matters. When managers update, and agenda items dominate one-on-ones, team members learn that their concerns and questions are secondary — and they start bringing less of themselves to the conversation.

The most important practice in a one-on-one is asking what the team member is working to figure out, not just what they are working on. "What's the hardest thing you're dealing with right now?" is a more generative question than "What did you accomplish this week?" The former invites genuine engagement with difficulty; the latter invites the construction of a positive narrative. A manager who consistently creates space for team members to name what is hard — without that naming taking the problem back —

builds the trust that makes performance feedback land more effectively when needed.

One-on-ones are also the right setting for ongoing developmental feedback — not the annual review, and not the spontaneous correction in a team meeting. When a manager observes something in the previous two weeks that should be reinforced or adjusted, the one-on-one is the moment to address it specifically: "I want to share something I noticed in Tuesday's stakeholder call that I think is worth discussing." This keeps feedback proximate to the behavior it references, substantially increasing its impact and retention. Feedback delivered six months after the fact, in an annual review, is nearly impossible to act on effectively because the team member's memory of the event and their emotional state at the time are too distant to generate actionable reflection.

Documenting one-on-one conversations — even briefly, in a shared note or a simple tracking tool — provides the raw material for accurate, specific performance reviews. A manager who has been capturing the topics and outcomes of 24 biweekly one-on-ones over the year has a rich, current record of the team members' challenges, growth, accomplishments, and development needs. That record transforms the performance review from an exercise in memory reconstruction into a synthesis of documented observations — and it shows the team member that the manager has been paying genuine attention throughout the year, not just at evaluation time.

6.11 Goal Conversations Throughout the Year

The calendar event most likely to be underinvested, rescheduled, and eventually eliminated from a busy manager's week is the goal check-in. This is precisely backward. Goal check-ins are the mechanism through which all the front-end investment in goal-setting yields an actual return. A team that sets clear goals and then does not revisit them regularly is a team that has done the administrative work of goal-setting without realizing any of its motivational or alignment benefits.

A quarterly goal review conversation — distinct from the weekly one-on-one and more structured than a casual check-in — is the right cadence for most teams. This conversation covers three questions: What progress have you made toward each goal since our last review? What obstacles are in the way, and what do you need to address them? Do any of the goals need to be updated, given changes in team or organizational priorities? The last question is the one most managers resist, because updating goals feels like lowering standards. In reality, a goal that no longer reflects current reality is not a standard — it is an administrative artifact that distracts from the work that actually matters.

These conversations also reveal important information about the quality of the original goal-setting process. Suppose a team member has made little progress on a goal and cannot articulate a clear reason why; the goal was probably not specific enough to generate an actionable work plan. If a team member has easily achieved a goal by the halfway point of the period, it was probably not sufficiently challenging. Both are useful signals for improving the quality of goal-setting in the next cycle, and neither should be treated with blame — they are diagnostic information, not indictments.

Connecting goal review conversations to recognition practice amplifies their impact. When a manager reviews a team member's progress on a Key Result and finds that the target has been met or exceeded, the goal review conversation is also a recognition conversation: "You committed to reducing report generation time by fifty percent, and you've exceeded that — you're at sixty-three percent. I want to acknowledge that, specifically because it required a significant process redesign effort that I know was not in the original project scope." Recognition delivered in the context of goal achievement is the most specific, credible, and motivating form of recognition available — it directly connects the team member's effort to the outcomes the team values most.

For managers leading hybrid or fully remote teams, goal visibility tools take on added importance. When the natural information exchange of a shared physical workspace is absent, the shared goal dashboard becomes the primary mechanism through which team members understand the collective picture of where the team stands relative to its commitments. Investing in a well-maintained, accessible goal-tracking system — whether on a dedicated platform like Lattice or Workday, or in a well-designed shared document — ensures that goal awareness is not dependent on the frequency of direct communication between a manager and a team member. Every person on the team should be able to answer, at any point in the quarter, exactly where each of their goals stands and exactly what they are focused on this week to advance them. That level of operational clarity, maintained consistently, is the foundation of a high-performance team culture.

6.12 Manager's Performance Management Checklist

At the beginning of each performance period, confirm the following. Goals are written in the OKR or SMART format — specific, measurable, time-bound, and aligned with organizational priorities. Each team member's goals have been discussed with them, not simply handed to them. The criteria for success at each performance level are explicit and documented, not left to end-of-year interpretation. Team member development goals are captured alongside performance goals — not treated as separate or secondary. You have confirmed that each team member understands not just their goals but why those goals matter to the team and the organization.

Throughout each performance period, verify the following regularly. Check-ins include explicit reviews of goal progress, not just project status updates. Positive performance patterns are being recognized specifically and promptly — not saved for the annual review. Developmental observations are shared in real time, allowing team members to adjust as the work progresses. AI-powered analytics tools are being reviewed at least monthly for early warning signals on engagement, well-being, and performance trends. Goal updates are documented when priorities shift — outdated goals are corrected rather than evaluated.

During each performance review, ensure the following. The team member's self-assessment has been collected and reviewed before the manager's assessment is shared. Every piece of feedback is grounded in a specific example — no generic ratings without behavioral evidence. The review covers not just what happened but what it means for the next period's goals and development plan. Underperformance has been named directly and specifically if it exists, with a concrete improvement path outlined. The conversation ends with mutual commitment to the next period's goals and development priorities — both parties leave with the same understanding of what success looks like going forward.

6.13   The Performance Partnership

Performance management is not something a manager does to a team member. It is something a manager does with them — a partnership built on a shared understanding of goals, an honest exchange of observations, and a mutual commitment to development. When that partnership is working, the annual review becomes a formality rather than a revelation. Team members know where they stand because they have been consistently and specifically told throughout the year. Goals are current because they have been updated when reality changed. Development is happening because coaching conversations have been ongoing rather than annual.

The AI tools available today — from analytics dashboards to goal-tracking platforms to sentiment analysis — make the informational infrastructure of performance management substantially more manageable. They reduce the time spent on data assembly and increase the time available for the human conversations that performance management ultimately depends on. They surface patterns that would be invisible to even the most attentive manager manually tracking a large team, and they do so in formats that make them immediately actionable.

But no technology changes the fundamental dynamic: people perform at their best when they know what is expected, understand why it matters, receive honest and timely feedback, and believe that their manager is invested in their success. Those conditions are created by a manager who sets clear goals, monitors progress with genuine attention, conducts honest reviews, and develops people with the same energy they bring to managing outputs. Build that practice consistently, and performance stops being something you manage after the fact and starts being something you lead from the front.

7 Hiring, Onboarding, and Building Your Dream Team

7.1 Scenario

The request came on a Tuesday: Elena's team had been approved to add two new positions, and her director wanted to move quickly. Her current team of five was stretched across three active projects, the existing members had been covering the gap left by a departure for nearly four months, and the business pressure to fill the roles was real and visible. Elena had never hired anyone before. She was handed a job description template from HR, told to "post it and start interviewing," and given a six-week timeline to extend offers.

She posted the jobs, collected resumes, and moved quickly. Within two weeks, she had twelve candidates in process. The interviews were conversational — she asked each candidate about their background, what they liked about their previous roles, and why they were interested in the position. She picked the two people she liked most from those conversations, made the offers, and congratulated herself on moving fast. Her director was pleased with the timeline.

Six months later, one of the hires had resigned after three months, citing a disconnect between what was described in the interview and the reality of the role. The other was still on the team but producing well below expectations — and Elena was now spending significant time coaching and correcting the work that, in her hiring process, she had assumed would be handled independently. The six-week timeline that had felt like an efficiency triumph was now generating months of remediation.

Elena's experience is common. Hiring is one of the highest-leverage activities a manager engages in — a great hire compounds a team's capabilities for years. In contrast, a poor hire consumes management time, team morale, and organizational resources long after the offer letter has been signed. And yet hiring is also one of the most commonly underprepared activities in management. This chapter provides the framework, the practices, and the AI tools that turn hiring from a reactive scramble into a disciplined, repeatable, and significantly more successful process.

7.2 Why Hiring Is the Highest-Leverage Activity

Every hire you make shapes your team's composition for the next several years. The right person in a role does not just complete work — they raise the ceiling of what the team can accomplish, bring perspectives that sharpen the team's thinking, model behaviors that spread through the team's culture, and, in many cases, develop into the next generation of leadership. The wrong person in a role does the opposite: creates drag on output quality, consumes disproportionate management attention, introduces friction into the team's dynamics, and often departs — at high cost and disruption — after a year or less.

Research on the cost of a bad hire consistently estimates that figure at one and a half to three times the role's annual compensation, when you account for recruitment costs, onboarding investment, lost productivity during the performance management and eventual replacement process, and the hidden costs borne by remaining team members who absorb the gap. For a manager leading a team of five to ten people, one consequential hiring mistake per year is a material drag on the team's performance and the manager's own leadership record.

The upstream investment that prevents this cost is the quality of the hiring process. Specifically: the precision of the role definition, the rigor of the interview structure, the deliberateness of the bias-reduction practices, and the quality of the onboarding experience that follows a successful offer. Each of these elements is within the manager's control, and each of them can be substantially improved with appropriate preparation and, increasingly, with AI tools that augment human judgment throughout the process.

Diagram 6.1: The Hiring Leverage Equation

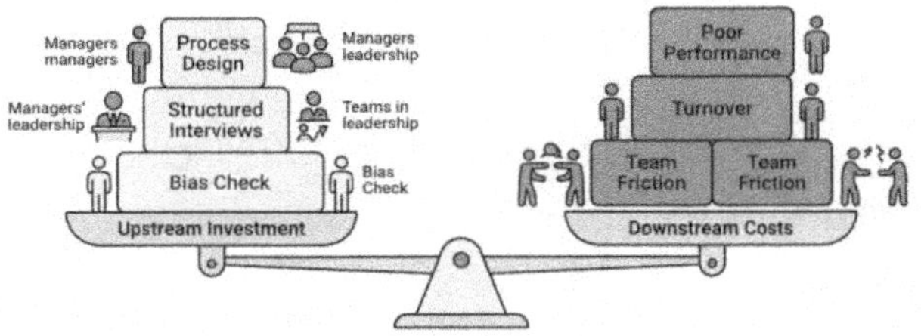

Effective hiring starts with intentional investment.

The AI dimension transforms the hiring equation in significant ways. AI tools can now screen hundreds of resumes in seconds, identify candidates whose profiles match the role's requirements, flag language patterns in job descriptions that may deter certain candidate populations, conduct preliminary assessments, and even generate structured interview guides. These capabilities do not eliminate the need for human judgment — and they introduce risks of their own, which this chapter will address directly — but they enable managers to operate a more thorough and consistent hiring process without a proportional increase in time investment.

7.3 Defining the Role You Actually Need

The single most important thing a manager can do before posting a position is to think carefully about what the role is actually intended to accomplish — not just what tasks the previous person in the role performed, but what capabilities the team needs, what problems the hire is expected to solve, and what success in the role looks like eighteen months after the person starts.

Most job descriptions are written by looking backward: they catalog the tasks the previous occupant performed, list the technical requirements that the job appeared to require, and append a generic list of soft skill phrases that HR has approved for use. The result is a description of a past role, not a future-oriented specification of a capability the team needs. This creates the hiring equivalent of looking for yesterday's fit — you find candidates who look like the last person, not the ones best suited for the next chapter of the team's work.

A more effective approach begins with what is sometimes called a success profile — a description of what the person in this role will have accomplished in their first year if the hire is successful, and what capabilities will have been demonstrated in the process. This profile answers three questions: What specific outcomes will this person produce? What competencies — not just skills, but combinations of knowledge, skill, and judgment — are required to produce those outcomes? And what experiences and background make it likely that a candidate has developed those competencies?

Diagram 6.2: Success Profile vs. Job Description

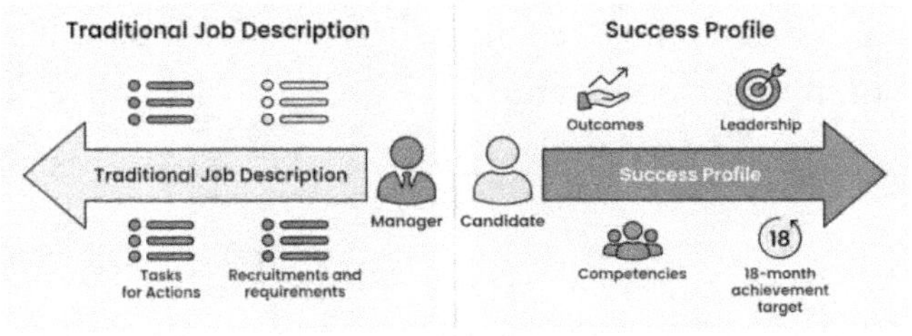

Caption: A Success Profile versus Job Description, being both outcomes-and-competencies joined for one.

Writing a compelling job description from the success profile is a separate skill — one where AI tools genuinely help. Modern hiring platforms and AI writing tools can analyze a job description draft for language patterns that inadvertently signal a preference for certain demographics. Research has documented that language emphasizing aggressive, competitive, and dominant qualities tends to deter women from applying, while language emphasizing collaboration and contribution tends to attract more diverse candidate pools. Tools like Textio and Ongig perform this analysis automatically, suggesting alternative phrasing that preserves the role's requirements while expanding the potential applicant pool.

The specificity of the job description also signals to candidates how seriously the organization takes the role. A generic job description that could describe a hundred different positions at a hundred different companies tells candidates very little about what they would actually be doing and why it matters. A specific, well-crafted description that articulates the team's mission, the role's concrete responsibilities, and the success criteria for the first year attracts candidates who have genuinely read it and self-selected based on a genuine fit, which is exactly the candidate screening mechanism you want operating before any human review begins.

7.4 Structured Interviewing That Predicts Success

Unstructured interviews — conversational exchanges in which the interviewer asks whatever comes to mind as the conversation goes on — are among the least predictive tools in the hiring process. Decades of research converge on a consistent finding: unstructured interviews have a validity coefficient of roughly 0.38 for predicting job performance, meaning they explain less than fifteen percent of the variance in how well a candidate will actually do the job. Despite this, unstructured interviews remain the dominant practice in organizations worldwide because they feel productive and revealing, and they confirm the interviewer's biases in ways that create false confidence.

Structured interviews substantially improve predictive validity — meta-analyses estimate that high-quality structured interviews achieve validity coefficients in the range of 0.51 to 0.63, roughly four times that of unstructured conversation. The mechanism is straightforward: when every candidate is asked the same questions, in the same sequence, and evaluated against the same criteria, the interview becomes a measurement instrument rather than a social encounter. The manager's job shifts from "getting a read" on the candidate to gathering evidence on a defined set of competencies.

The two most validated structured interview techniques are behavioral interviews and situational interviews. Behavioral interviews ask about past behavior on the premise that past behavior is the best predictor of future behavior: "Tell me about a time when you had to deliver a project with an incomplete set of requirements. What did you do, and what was the outcome?" Situational interviews present hypothetical scenarios that reflect the actual challenges of the role and ask candidates how they would handle them: "If you joined our team and discovered in your first week that the team had misunderstood a critical client deliverable for the past month, how would you approach that situation?"

Both techniques require pre-defined rating criteria — what a strong answer looks like and what a weak answer looks like — developed before the interviews begin. Without these criteria, interviewers default to impression-based evaluation, which is precisely what structured interviewing is designed to prevent.

Diagram 6.3: Structured Interview Design

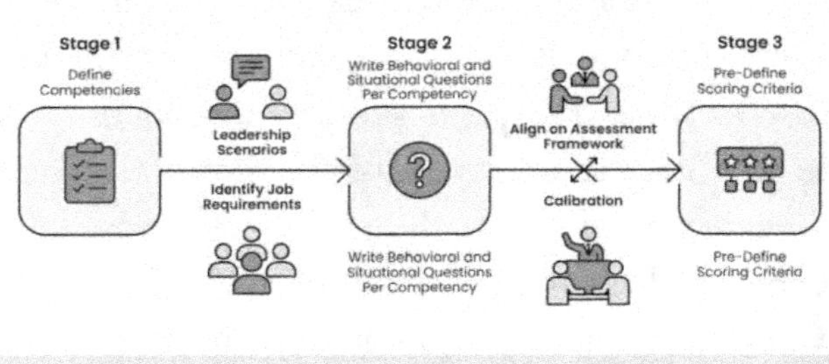

Building an effective interview panel means distributing coverage of different competencies among interviewers, then running a structured debrief in which each interviewer shares evidence-based assessments — not overall impressions — on the competencies they were assigned to evaluate. The hiring manager's role in this debrief is to facilitate the review of evidence rather than advocate for their preferred candidate. When every panel member has gathered specific behavioral evidence and evaluated it against pre-defined criteria, the debrief becomes a genuine calibration exercise rather than a negotiation between competing intuitions.

AI tools are increasingly useful in structured interview design. Generative AI can produce a first draft of behavioral and situational questions calibrated to specific competencies, saving the hiring manager time by writing questions from scratch while ensuring they align with the role's actual requirements. The manager's job is to review and refine those questions, not generate them. This is an appropriate application of AI augmentation: it accelerates a task while keeping the manager's judgment in the loop for the decisions that matter most.

7.5 Reducing Bias in the Hiring Process

Hiring bias is not an edge case or an aberration — it is the default condition of any hiring process that does not actively work to counteract it. Human cognition is built for pattern recognition and rapid judgment, and those capabilities lead to systematic errors when applied to evaluating candidates. Affinity bias leads interviewers to favor candidates who remind them of themselves. The halo effect causes a strong first impression to color subsequent evaluation. Recency bias inflates the weight placed on the last few exchanges of an interview. Attribution bias leads to different interpretations of the same behavior depending on the person's demographic characteristics. These biases operate largely below the interviewer's conscious awareness, which is why good intentions are not a sufficient guard against them.

The structural antidote is process design that reduces bias by reducing the role of unconstrained judgment in high-stakes decisions. Structured interviews, calibrated scoring criteria, diverse interview panels, and consensus-based hiring decisions are mechanisms that distribute evaluation across multiple perspectives and anchor judgments to pre-defined standards rather than subjective impressions. No process eliminates bias — but these structural protections meaningfully reduce its influence.

Anonymous review — evaluating resumes with identifying information removed — is among the most researched bias-reduction practices. Audit studies have consistently found that resumes with names associated with minority groups receive fewer callbacks than identical resumes with names associated with majority groups, even when all qualifications are held constant. Removing names, graduation years, and other identity markers from the resume review stage reduces this effect. It ensures that initial screening decisions are based on qualifications rather than on demographic inference.

AI-powered resume screening tools introduce a specific bias risk that managers need to understand explicitly. These tools learn patterns from historical hiring data, which means that if historical hiring practices were biased, the AI will learn and replicate those patterns unless corrective measures are built into the model. Several high-profile cases have documented AI recruiting tools that downgraded resumes from candidates who attended women's colleges or who included words associated with female professional networks, because historical data showed those candidates were less frequently hired. Using AI in resume screening without examining the model's training data and validation outcomes for demographic bias is not a bias-reduction measure — it is a bias laundering mechanism.

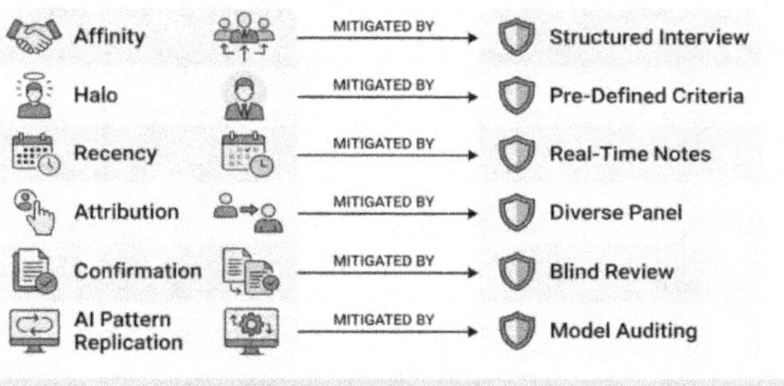

The manager's responsibility in bias reduction extends beyond their own interview behavior to the entire process they oversee. This means choosing AI tools whose vendors can demonstrate bias auditing, insisting on diverse interview panels even when it is logistically inconvenient, and examining aggregate hiring data periodically to check whether the process is producing outcomes that are demographically consistent with the candidate pool. This examination does not require sophisticated analytics — it requires the willingness to look at the data and take the results seriously.

7.6 AI-Powered Recruiting and Assessment Tools

The AI tooling available to managers in the recruiting process has expanded dramatically in the last several years, and understanding both the capabilities and the limitations of these tools is essential to using them responsibly.

Applicant tracking systems like Greenhouse, Lever, and Workday Recruiting have incorporated AI features that go well beyond basic resume management. These platforms now offer automated screening that uses machine learning to analyze resume content and work history and rank candidates by predicted fit with the role's requirements. They offer candidate matching that identifies passive candidates in a talent network who may not have applied but whose profiles are highly relevant. And they offer structured feedback collection and calibration tools that make the interview panel debrief significantly more rigorous.

Pre-employment assessment platforms — HireVue, Pymetrics, Codility, and similar tools — extend AI capabilities into the evaluation phase. Video interview analysis tools analyze verbal and nonverbal communication patterns in recorded interviews to assess competencies such as clarity of communication, structured thinking, and interpersonal engagement. Skills-based testing platforms assess specific technical capabilities — coding assessments, data analysis exercises, writing samples — in standardized formats that allow objective comparison across candidates. These assessments are more predictive of role-specific technical skills than traditional interviews and significantly reduce the time investment required to evaluate a large candidate pool.

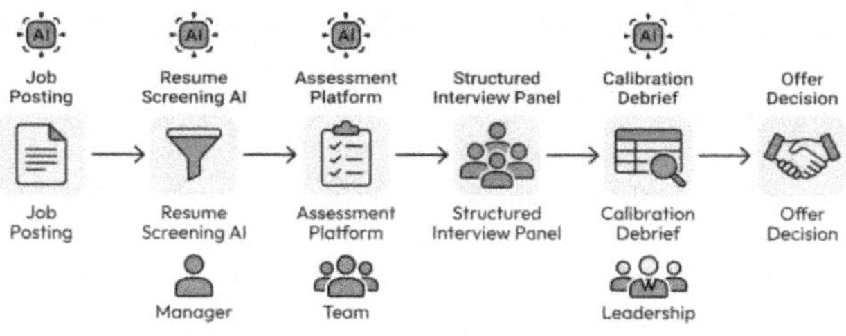

The limitations of these tools are as important as their capabilities. Video interview analysis is among the most controversial AI applications in hiring: the evidence for its validity in predicting job performance is significantly weaker than vendors typically represent, and there are documented concerns about differential performance across demographic groups for facial expression and vocal pattern analysis. Skills assessments are more defensible — a coding test that measures whether a candidate can write functional code has face validity that is hard to dispute — but they should be designed to reflect actual work conditions rather than artificial puzzle solving challenges.

The responsible use of AI in recruiting follows a principle that this book returns to throughout: AI should narrow the field and inform the decision, not make the decision. AI resume screening that surfaces the 50 most relevant candidates from a pool of 300 makes the manager's job more manageable without removing human judgment from the evaluation. An AI that makes hiring recommendations without human review has entered a territory where accountability is unclear, and the risk of systematic error is high. Every AI-assisted hiring decision should be explainable — the manager should be able to articulate, in human terms, why this candidate was selected — not merely attributed to an algorithm's output.

7.7 Designing an Onboarding Experience That Works

Hiring and onboarding are a single process, not two sequential events. The quality of the onboarding experience directly determines how quickly a new hire becomes productive, how strongly they connect to the team's culture, and whether the match that looked compelling in the interview process is confirmed or undermined in the first months of actual work. Research consistently finds that structured, well-designed onboarding programs improve retention by up to 82% and productivity timelines by up to 70%, compared to ad hoc approaches that leave new hires to figure things out by observation and trial.

The most common onboarding failure is the information dump. The new hire arrives on Monday, spends the first day in HR paperwork and compliance training, receives access to fourteen internal systems, is introduced to forty people whose names they will not remember by Wednesday, and is expected to begin contributing meaningfully within a few weeks. This approach treats onboarding as a logistics exercise rather than a development experience. The new hire is overwhelmed, undercontextualized, and left without a clear picture of what they should be doing or why any of the information they received in their first week is relevant to their actual work.

Effective onboarding is structured around three dimensions that build sequentially: connection, context, and contribution. Connection comes first — the new hire needs to feel that they belong, that their manager is invested in their success, and that they have at least one or two genuine points of relationship within the team. Without connection, context is fragile, and contribution is anxious. Context follows — the new hire needs to understand the team's mission, its work methods, its current priorities, the stakeholder landscape, and the informal rules of operation that never appear in any documented process. With connection and context established, the pathway to contribution becomes navigable.

Diagram 6.6: The Three-Dimension Onboarding Framework

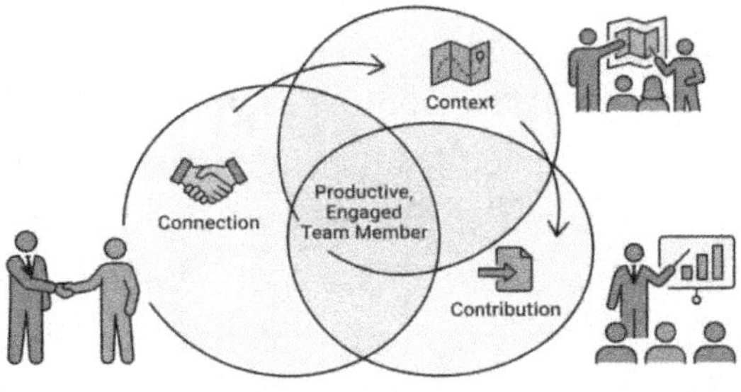

Integrating Connection, Context, and Contribution accelerates new hire productivity and engagement.

A thirty-sixty-ninety-day onboarding plan operationalizes this framework. The first thirty days are primarily about connection and orientation: meet the key stakeholders, understand the team's current priorities, observe rather than redesign, and identify one or two small early wins that demonstrate competence without requiring full context. The second thirty days shift toward active learning and early contribution: take on a defined project with appropriate support, develop a deeper understanding of the team's methods and tools, and begin identifying the areas where the new hire's specific strengths can add distinctive value. The third thirty days are about sustained contribution and integration: own meaningful work, provide and receive feedback as a full member of the team, and articulate what success looks like in the role over the next quarter.

AI tools can accelerate onboarding in specific ways. AI-powered knowledge management platforms like Notion AI, Guru, and Confluence AI give new hires searchable access to institutional knowledge that would otherwise require months of informal learning from colleagues. AI-generated onboarding guides can be customized to a specific role and team context, synthesizing information from multiple internal sources into a structured learning path. AI meeting transcription and summarization tools allow new hires to catch up on important context from meetings they missed without requiring colleagues to spend time providing handcrafted summaries. These tools significantly compress the context-building phase of onboarding, directly accelerating the pathway to productive contribution.

7.8 Building Team Diversity and Complementary Skills

A team where everyone thinks the same way, shares the same background, and approaches problems with the same tools is vulnerable to blind spots. Diversity in teams —

demographic, cognitive, experiential, and disciplinary — is not primarily a compliance imperative or a reputational benefit, though it is both. It is a performance advantage, supported by substantial research, that produces better decision quality, more creative problem-solving, and more robust risk assessment than homogeneous teams can achieve.

The mechanism is not mysterious: when a team comprises people with genuinely different perspectives and experiences, the range of factors considered in any decision is greater, the range of anticipated failure modes is broader, and the probability that a groupthink error will go unchallenged is lower. Teams that feel comfortable challenging each other's assumptions — which requires psychological safety — derive more value from their diversity than teams where different perspectives are present but not actively engaged.

Building complementary skills is a related but distinct goal. Skill complementarity means that the team's collective capabilities cover the full range of work the team needs to perform, without so much redundancy that team members are competing for the same tasks. A team with five strong individual contributors and no one with project coordination capabilities will consistently miss on delivery. A team with excellent technical depth but no one who can translate that depth into stakeholder communication will consistently struggle with organizational influence.

When you approach a hire, consider the team's current skill map before writing the job description. Identify the gaps — not just the vacancy left by a departed team member, but the capability gaps that have been limiting the team's performance regardless of that vacancy. Sometimes the answer is to hire for a profile significantly different from the previous occupant of the role, because the team's needs have evolved. Sometimes it confirms the original profile. Either way, the analysis makes the decision deliberate rather than reflexive.

Identifying and bridging critical skill gaps enhances overall team performance and strategic capabilities.

7.9 The Manager's Role in the First 90 Days

Their manager's behavior primarily shapes a new hire's experience in the first ninety days. The decisions you make about how much time to invest, what experiences to create, and what signals to send about the team's culture and your own leadership style will determine whether the person you hired remains confident and engaged or becomes uncertain and disengaged.

The most common manager mistake during new-hire onboarding is unavailability. Genuinely busy managers — which is most managers — often treat the new hire's first weeks as a period when the team will take care of the newcomer informally. At the same time, the manager attends to existing priorities. This approach underestimates how much a new hire's experience depends on direct engagement with the manager in the early period. The new hire is reading every interaction with the manager for signals about expectations, standards, and relationship quality. Manager unavailability reads as disinterest, which reads as low confidence in the hire — exactly the opposite of the psychological foundation the new hire needs to perform.

Weekly one-on-ones during the first ninety days are non-negotiable. These conversations serve multiple purposes simultaneously: they give the manager real-time visibility into how the new hire is experiencing the onboarding process; they give the new hire a reliable space to ask questions, surface concerns, and receive guidance; and they build the relationship that will make difficult conversations easier when they inevitably arise. The first 90 days are also the window when the new hire is most willing to ask basic questions and most open to receiving guidance—take advantage of it.

The thirty-sixty-ninety-day plan should be co-created with the new hire, not presented as a fixed requirement. Involving the new hire in defining their own early milestones builds ownership. It creates a more accurate plan, since the new hire often has insights about their own learning style and development priorities that the manager does not yet know. It also communicates that the manager sees the new hire as a partner in their own development, not a subject of an onboarding protocol.

The end of the first ninety days is an appropriate moment for an explicit "how is this going" conversation — not a formal performance review, but a candid two-way exchange about whether the role is what both parties expected, what is working, what needs adjustment, and what the priorities should be for the next quarter. This conversation is significantly easier and more productive than the same conversation at the one-year mark, when more time has passed without course correction and when perceptions on both sides are more entrenched.

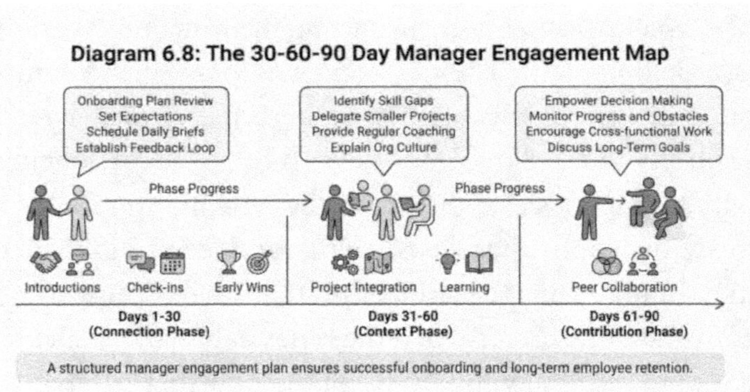

7.10   The Candidate Experience Is a Leadership Signal

Every interaction a candidate has with your organization — from the moment they read the job posting to the moment they receive the offer or decline — is an experience that shapes their perception of what working for you would be like. This is not primarily a concern of recruitment marketing. It is a leadership signal. The way you and your team treat candidates during the hiring process communicates your team's values, your respect for other people's time, and your operational competence in ways that are impossible to fake over several rounds of interviews.

The most damaging candidate experience failures are also the most preventable. Disappearing after an interview — leaving candidates in silence for two or three weeks with no update — tells them that communication is not a priority in your organization. Arriving at a video interview five minutes late, unprepared, and clearly not having reviewed the candidate's resume tells them that their time is less valuable than yours. Describing the role in significantly more appealing terms than the reality creates a mismatch that leads to turnover within the first year, because the hire's expectations were not calibrated to the job's experience.

The positive version of the candidate experience is equally communicative. When a candidate leaves an interview thinking "that manager asked thoughtful questions, was fully present, and gave me a clear picture of what the role actually involves," they carry that impression through the remainder of the process. If they receive an offer, they are more likely to accept it — not just because the compensation is competitive, but also because the manager has given them confidence that joining the team would be a good decision. If they receive a decline, they are more likely to remain positive about the organization and potentially refer other candidates, because the experience treated them with respect even when the outcome was not what they hoped for.

Speed and communication during the hiring process are dimensions that managers control more than they often recognize. Committing to a maximum response time after each interview round — communicating that commitment to candidates — and then honoring it is a differentiating practice in a world where candidates regularly wait weeks without updates. It requires discipline and prioritization on the manager's part. Still, it is discipline that pays dividends in offer acceptance rates and in the caliber of candidates who stay engaged with the process rather than accepting alternative offers while waiting to hear back.

The candidate experience is also a team development opportunity. When you include team members in the interview panel, you are not just collecting additional evaluation perspectives — you are giving team members experience in structured hiring, practice in evidence-based assessment, and visibility into the qualities you value in a new colleague. Debriefing with the panel after interviews, discussing what each interviewer observed, and making the hiring decision in a way that is transparent to the team builds the collective hiring capability that will serve the team for years beyond any individual hire. A team that understands how hiring decisions are made can participate more constructively in the next one.

7.11 Manager's Hiring and Onboarding Checklist

Before posting the role, verify the following. A success profile has been written — not just a task list — describing what the hire will have accomplished in twelve months and what competencies they will have demonstrated. The team's current skill map has been reviewed, and the hire's role in addressing skill gaps has been explicitly taken into account. The job description has been reviewed for biased language using an AI analysis tool or a structured bias checklist. Interview questions have been written in advance, tied to specific competencies, with pre-defined scoring criteria for strong and weak answers. The interview panel has been selected to provide diverse perspectives and to cover competencies across the organization.

During the interview process, confirm the following. Every candidate is asked the same questions in the same sequence. Interviewers are taking notes on specific behavioral evidence, not impressions. The interview debrief is structured around competency ratings with evidence, not overall preferences. AI screening and assessment tools are being used to narrow the field, not to make the final decision. Bias-check practices — anonymous resume review, calibrated scoring — are active at every stage.

For every new hire, ensure the following. A thirty-six to ninety-day plan has been written, discussed with the new hire, and includes clear milestones for each phase. Weekly one-on-ones are scheduled for the first ninety days and are consistently held. Key introductions — stakeholders, cross-functional partners, team members — are arranged in the first two weeks rather than left to emerge organically. Access to tools, systems, and documentation is ready before the first day. AI-powered knowledge management access is configured so the new hire can self-serve the institutional context. A ninety-day "how is this going" conversation is scheduled at the start, not added retroactively.

7.12   Hiring as Team Architecture

Every hire is an architectural decision. You are not just filling a vacancy — you are choosing the person who will shape the team's capability, culture, and trajectory for the next several years. That decision deserves the same deliberateness you would bring to any significant structural choice.

The process described in this chapter — role definition grounded in a success profile, structured interviewing with calibrated scoring, active bias reduction, responsible use of AI tools, and deliberate onboarding designed around connection before contribution — is more demanding than posting a job and interviewing whoever applies. It is also substantially more likely to produce the team you are trying to build.

The AI tools available today make it possible to execute this process more efficiently than was feasible a decade ago. Resume screening at scale, skills matching, assessment platforms, onboarding, and knowledge management — each of these applications reduces the friction associated with a disciplined hiring process and increases the volume of useful information available to the manager's judgment. They do not remove the need for judgment; they amplify its quality and reach.

The cumulative effect of hiring well — not once, but consistently over years — is a team whose capability compounds. Each strong hire raises the average, attracts other strong candidates by the quality of the environment, and builds the organizational credibility that gives you access to better roles and more resources. Conversely, each consequential hiring mistake is an anchor that limits the team's trajectory and consumes management energy that should be spent on leadership, not remediation. Invest in the process. The returns are among the highest available to a manager.

8 Coaching, Feedback, and Developing Talent

8.1 A Scenario Worth Sitting With

Three months into her role as a software engineering manager, Priya had a strong team on paper. Her six direct reports were experienced, technically sharp, and generally hit their sprint commitments. Yet something was off. In their one-on-ones, conversations stayed surface-level. People reported on tasks rather than talking about growth, challenges, or what was energizing or draining them. Two team members were quietly applying to other companies. A third had become disengaged — still producing, but operating well below the ceiling everyone on the team could sense was there. Priya had plenty of feedback she wanted to give — some of it critical — but every time she started to formulate it, she felt a pull to soften it so much that it would lose its meaning, or she worried the person would shut down and stop bringing real problems to her at all. She told herself she would get to it next week. Next week came. She told herself, next sprint.

What Priya was experiencing is one of the most common and most consequential challenges facing new managers: the gap between having a coaching mindset and actually deploying it. Feedback was available. Development conversations were possible. The structural tools — one-on-ones, performance reviews, individual development plans — were all in place. But nobody had ever taught Priya the mechanics of the coaching conversation, how to structure a development-oriented one-on-one, or how to deliver feedback in a way that would be heard rather than defended against. She defaulted, as most new managers do, to management by information exchange — project updates, dependency checks, status confirmations — when what her team needed was management by growth. The irony is that Priya was working harder than ever while making less of her most important contribution.

This chapter closes that gap. It gives you a complete operating framework for the manager as coach — covering how to give feedback that actually lands, how to run one-on-ones that develop people rather than update you, and how to build individual development plans that your team members actually believe in and pursue with genuine energy. It also explores how AI tools are beginning to change the coaching landscape — from surfacing skill gaps before they become performance problems to personalizing learning recommendations in ways that would have been logistically impossible even five years ago. Whether you are managing a team of three or thirty, developing talent is the highest-leverage work you will do. Every other management contribution has a ceiling set by your team's capabilities. Raising those capabilities is how you raise the ceiling.

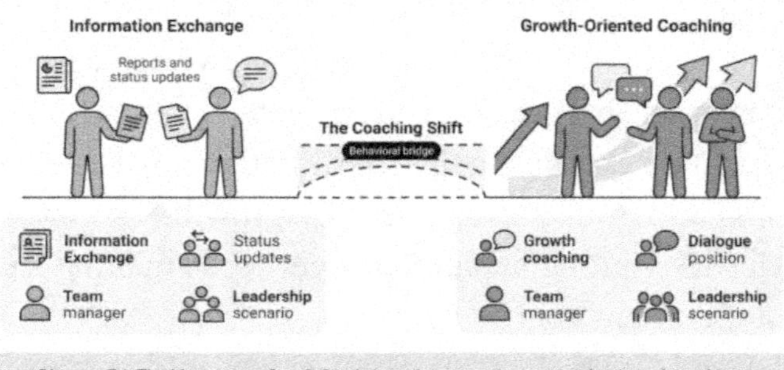

Diagram 7.1: The Manager-as-Coach Gap is in tantint sumention on growth-oriented coaching.

8.2 Why This Matters

Managers who coach outperform managers who supervise. The evidence across industries and organizational types is consistent and compelling: teams with coaching managers — those who actively invest in developing their people's capabilities — show higher engagement scores, lower voluntary turnover, faster skill acquisition, stronger cross-functional collaboration, and better business outcomes across virtually every measured dimension. This is not a soft observation made by management consultants who benefit from selling coaching programs. It is a quantifiable pattern that reliably appears in data from organizations that have studied it seriously. Gallup's landmark research on the manager's impact found that the manager accounts for at least seventy percent of the variance in team engagement. That variance is not explained primarily by the manager's technical expertise, their domain knowledge, or even their strategic intelligence. It is explained by whether the manager treats people as resources to be deployed to current tasks or as capabilities to be developed over the span of their careers.

From a mission-alignment perspective, coaching is not a luxury activity you do after your real work is done. Coaching is the real work. Every manager operates with a finite talent budget — the combined capability of the people on their team at any given point in time. Managers who coach expand that budget over time. Every person they develop adds to the team's aggregate output ceiling. Managers who do not invest in development see that the budget erodes slowly, as stagnation, disengagement, and eventually turnover diminish the team's collective capability. As AI tools automate routine tasks and raise the baseline expectation for what a skilled professional can produce, the gap between a developing employee and a stagnant one will only widen. The team that is learning continuously will compound its advantage. The team that is merely executing will find itself progressively outpaced by both human and AI competition.

Feedback, in particular, yields an outsized return on investment when applied with skill and consistency. Done well, a single, precise, behaviorally grounded piece of feedback can redirect a career trajectory, prevent a costly mistake from repeating, or unlock performance that was being suppressed by a blind spot the person was unaware of. Done poorly — vaguely, too late, too harshly, or without a relational foundation — feedback creates defensiveness, erodes trust, and drives people toward compliance rather than genuine growth. The difference between the two outcomes often has nothing to do with the content of the feedback itself. It has everything to do with the method, timing, relational context, and emotional intelligence with which it is delivered. This chapter gives you the tools to be on the right side of that difference consistently.

One additional dimension warrants explicit naming: the connection between coaching quality and team retention. In a labor market where skilled professionals have options, people do not stay in jobs where they feel invisible, stagnant, or undervalued. They stay where they feel seen, challenged in ways that matter, and supported by a manager who is genuinely invested in their growth. Retention is not primarily a compensation problem. Research from multiple sources consistently finds that the relationship with the direct manager — the quality of feedback, the sense of development opportunity, the feeling of being known and valued — is a more powerful predictor of voluntary turnover than salary in most professional contexts. A manager who develops the coaching skills in this chapter is not just improving team performance. They are protecting one of their organization's most expensive assets from walking out the door.

8.3 The Coaching Mindset — Asking vs. Telling

The most important conceptual shift in becoming an effective coaching manager is understanding the difference between a telling orientation and an asking orientation — and developing the judgment to know which one the moment calls for. A telling orientation means you treat your primary role as that of a transmitter of answers, decisions, and direction. You have the experience, the organizational knowledge, the technical expertise, and the perspective, and your job is to efficiently transfer those assets to your team so that they can execute correctly. An asking orientation means you treat your primary role as an activator of thinking, self-discovery, and intrinsic motivation. You believe the person in front of you has more capacity, insight, and capability than they are currently accessing, and your job is to create the conditions under which they access it.

Both orientations have their place, and the skilled coaching manager moves fluidly between them. There are absolutely moments when your team needs your answer, your decision, your clear direction — a production outage at eleven PM, a regulatory deadline with immediate legal implications, a new hire's first day when they need orientation rather than Socratic inquiry. Coaching your way through a crisis is not wisdom; it is negligence. The telling orientation is appropriate when the stakes are high, the time is short, the person genuinely lacks the knowledge or context to find their way through the problem independently, or when organizational accountability requires a definitive managerial decision. Knowing when to switch back to telling is as important as knowing how to ask well.

That said, most experienced managers err so far toward telling that they deprive their teams of the developmental opportunities that asking creates. When you always answer the question, you are always the answer. When you help someone find their own answer, they own it. They remember it. They build on it. They develop the capacity to generate it independently next time. This is the compounding return of the asking orientation: not just one problem solved, but one problem solver developed. The manager who consistently asks rather than tells is building a team of thinkers, not a team of implementers waiting for direction.

The GROW model, originally developed by Sir John Whitmore and refined over decades of executive coaching practice, remains one of the most practical frameworks for structuring a coaching conversation. GROW stands for Goal, Reality, Options, and Will. In a GROW-structured coaching conversation, you begin by helping the person clarify what they are trying to achieve — the specific outcome they want from this conversation or situation (Goal). You then help them build an honest assessment of where they currently are relative to that goal, without judgment or premature advice (Reality). Next, you explore the range of options available to them — brainstorming broadly and generatively before narrowing to commitments (Options). Finally, you establish what specific action the person will take, by when, and how they will know they have succeeded (Will). The manager's role throughout is to ask powerful questions that open up thinking: What matters most to you about this? What have you already tried? What else might be possible that you haven't considered? What would have to be true for that option to work? What do you most need from me right now?

The GROW model works because it respects a fundamental truth about adult development: people pursue most vigorously what they themselves have reasoned their way to. The solution that the coach tells you to implement lives in the coach's head. The solution that you construct through your own thinking, with the coach's questions as scaffolding, lives in your own understanding. You know why it makes sense. You know the trade-offs you have already considered. You know the contingencies you have built in. When implementation becomes difficult — and it almost always does — you have the internal resources to adapt because you built the solution, rather than merely executing someone else's blueprint.

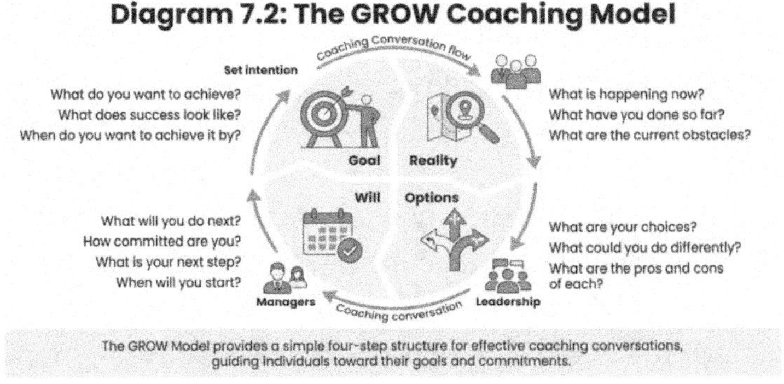

8.4 Giving Feedback That Lands — The SBI Model and Beyond

The SBI model — Situation, Behavior, Impact — is the gold standard framework for delivering behaviorally specific, defensible, and development-oriented feedback. The Center for Creative Leadership developed it and has been validated across cultures, industries, and organizational levels, consistently demonstrating its effectiveness over alternative approaches. The model works because it forces you to anchor feedback in observable, specific, verifiable reality rather than in interpretation, judgment, or character assessment. Instead of making evaluative pronouncements about the person ('you're too aggressive in meetings,' 'you're not a team player,' 'you need to be more executive-ready'), SBI requires you to describe what you actually observed in

188

a specific situation, what behavior you saw, and what effect that behavior produced.

Situation establishes the context: when and where did this happen? 'In our sprint review on Tuesday morning' or 'when you presented the Q3 forecast to the leadership team last Wednesday' are Situation statements. They are specific enough to be unambiguous and recent enough to be relevant. Behavior describes the observable action — what the person said, did, or did not do — without inference about motivation, intent, or character: 'you immediately listed the technical reasons the delay was not an engineering problem' or 'you referenced three different data sources without clarifying which one should take precedence.' Impact describes the actual, observed effect on the team, the work, a stakeholder, or an outcome: 'three team members visibly disengaged from the rest of the conversation' or 'the client left the call without a clear understanding of the timeline, and followed up with two escalation emails.'

The contrast between SBI feedback and vague character feedback is stark and important. 'You get defensive when stakeholders challenge the team' is a character assessment that the recipient will almost certainly resist, because it invites argument ('I wasn't being defensive, I was being accurate') and provides no actionable guidance on what to change. The SBI version of the same feedback — 'In the sprint review on Tuesday, when the product manager raised the delay on feature three, you immediately listed the reasons it was not an engineering problem, and three team members visibly disengaged from the rest of the conversation' — is specific, observable, and grounded in reality. The recipient may initially feel defensive, but they cannot honestly deny what happened. They have a clear behavioral target to work with.

Two extensions of SBI substantially increase the framework's practical power. The first is adding a fourth element: Inquiry. After stating the Situation, Behavior, and Impact, you pause and invite the person to respond before you draw any conclusions or make any requests. The transition statement is: 'I wanted to share that observation and hear your perspective on it.' This shift from feedback as verdict to feedback as conversation changes the entire dynamic. The person may have context you lack — they may have received conflicting direction from another stakeholder, may have been responding to a pattern of behavior they have been waiting for an opportunity to address, or may have made a deliberate calculation that backfired. Their response is information, not just an emotional reaction to be managed. More importantly, engaging them as a thinking partner in analyzing their own behavior is far more likely to produce lasting behavioral change than presenting them with a judgment they feel compelled to resist.

The second extension is calibrating your feedback channel to the nature of the message. Positive, reinforcing feedback should be delivered as close in time to the observed behavior as possible — the reinforcement value degrades rapidly with delay — and should often be public: acknowledgment in the team standup, a Slack message in a visible channel, a genuine compliment in front of peers. The specificity matters even for positive feedback; 'great job on the presentation' is less valuable than 'the way you handled the CFO's unexpected question about margin assumptions — staying grounded, acknowledging what you didn't have immediately, and committing to a specific follow-up — was exactly the kind of executive presence this team needs more of.' Corrective or developmental feedback should almost always be given privately, promptly, and calmly. The manager who saves up six months of concerns for a performance review has not been giving feedback — they have been assembling a case. The goal of feedback is not to prosecute. It is to develop.

Frequency matters as much as quality. Research on behavioral change consistently shows that frequent, low-intensity feedback — brief, specific, conversational — produces better and more durable outcomes than infrequent, high-stakes feedback sessions. The formal performance review should never contain a surprise. If it does, it means the feedback channel has been closed for the entire review period. A team member who receives specific, behavioral feedback regularly — weekly if the relationship and workflow support it, at a minimum monthly — builds a much clearer and more accurate mental model of their own performance than one who receives feedback only in formal cycles. Building this cadence requires deliberate practice, but it quickly becomes natural.

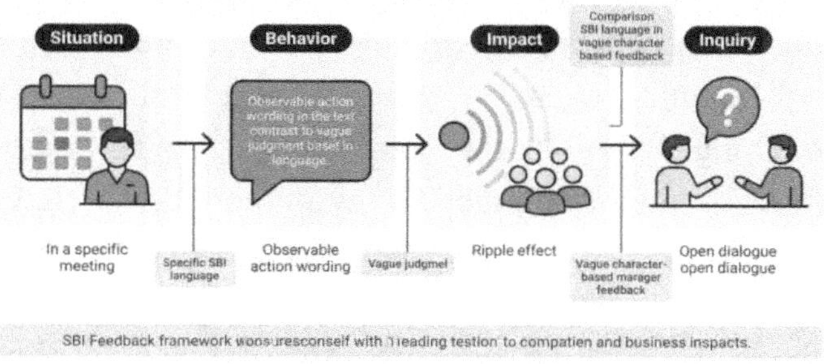

Diagram 7.3: SBI Feedback Framework

8.5 Creating a Culture of Continuous Feedback

One of the deepest structural problems with feedback in most organizations is that it flows primarily in one direction — downward from manager to direct report — and occurs infrequently. Performance reviews happen quarterly or annually. Skip-level feedback arrives rarely, if at all. Peer feedback, when it exists, is embedded in formal review cycles that most people treat as compliance exercises rather than development inputs. The result is a feedback desert: people go months without any meaningful, specific signal about how they are showing up, what is working, and what needs to change. This silence is not neutral — it is actively harmful, because in the absence of real feedback, people fill the gap with their own assumptions, which are often inaccurate and frequently pessimistic.

Building a continuous feedback culture starts with the manager's own modeling behavior. When you give feedback regularly — specifically, conversationally, and without drama — you normalize it as a routine feature of professional life rather than a high-stakes event to be dreaded. When you visibly seek feedback from your own team about your management — 'What's one thing I could do differently to make your work easier?' or 'Is there something I said in the team meeting today that landed differently than I intended?' — you model the behavior you are trying to instill. And when you receive feedback from your direct reports with visible, genuine curiosity rather than defensiveness, you demonstrate that the channel flows both ways and that using it is safe.

Several structural practices accelerate the formation of a continuous feedback culture. A brief, structured feedback exchange at the end of every sprint or project cycle creates a natural rhythm and connects feedback directly to work that is fresh in everyone's memory. A standing question in one-on-ones — 'What feedback do you have for me this week?' or 'Is there something you've been hesitant to bring up?' — creates a recurring invitation that, repeated consistently, gradually reduces the activation energy required to say something important. Celebrating specific instances of feedback well given and well received in team settings — 'I want to acknowledge that the direct conversation Keisha and Jamal had about the sprint planning process last week was exactly what this team's culture should look like' — publicly reinforces the norm and signals to everyone on the team what behavior is valued.

AI tools are increasingly capable of supporting feedback culture without replacing the relational and trust-based elements that make it work. Engagement platforms like Lattice, 15Five, and Culture Amp incorporate AI that can identify patterns in pulse survey data, flag when individual engagement scores drop significantly from their baseline, and prompt managers with suggested conversation topics based on what team members have signaled they are experiencing. Natural language processing capabilities in some communication platforms can identify when a team member has been consistently quiet in asynchronous channels or consistently absent from team decisions — early indicators of disengagement that might not be visible in direct conversations. These tools extend the manager's awareness beyond what is directly observable. Still, they do not replace the judgment required to interpret those signals or the relationship required to act on them effectively.

One additional practice that significantly strengthens the feedback culture is the post-project or post-retrospective feedback exchange structured explicitly around development rather than evaluation. At the end of a major initiative, asking each team member to share one thing they observed in each colleague that contributed to the team's success and one thing they would invite that colleague to consider for their own development creates a peer feedback channel that is both specific and contextual. Unlike peer feedback in performance reviews, which is aggregated and anonymized, this format allows for direct dialogue and follow-up questions. It also reinforces the norm that feedback is a gift rather than a judgment — something given to support someone's growth rather than to evaluate their worth.

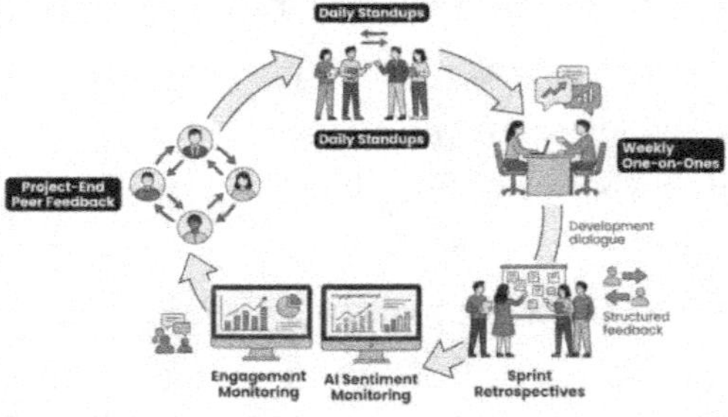

8.6 One-on-One Meetings That Drive Engagement and Growth

The one-on-one meeting is the most important recurring management instrument available to you. It is the dedicated space where the relationship between manager and direct report is built, maintained, and deepened over time. It is where career conversations happen organically rather than in the awkward annual review formality. It is where concerns surface before they have compounded into crises. It is where development progress is tracked, course corrected, and celebrated. It is where you demonstrate — through consistent attention, genuine curiosity, and reliable follow-through on commitments — that you take this person's growth seriously as a professional obligation and not just a managerial checklist item. And it is, more often than any other format, the place where the strongest team members either recommit to the work or begin the quiet process of deciding to leave.

Given all of this, the chronic undermanagement of one-on-ones in most organizations is remarkable. They are canceled when schedules get busy, which signals that the team member's time and growth matter less than the meeting that bumped them. They are kept superficial, reduced to project status exchanges that could have been Slack messages. They are conducted back-to-back with seven other one-on-ones on the same day, by a manager whose attention quality has degraded substantially by the fourth conversation. Or they are treated as manager agenda first sessions, where the direct report gets fifteen minutes to answer the manager's update questions before the manager moves on to their own talking points. Every one of these patterns sends a message about priorities. None of those messages supports engagement or retention.

The most effective one-on-ones follow a consistent structural logic while remaining flexible enough to accommodate what the person actually needs on a given week. A useful default architecture has three parts. The first ten to fifteen minutes belong entirely to the team member. They set the agenda. They bring what is top of mind — a concern they have been carrying, an idea they want to think through, a professional challenge they need a sounding board for, or a personal circumstance that is affecting their work. Your role in this segment is to listen actively, ask clarifying questions that deepen your understanding, and resist the gravitational pull toward your own agenda. The middle portion — roughly fifteen to twenty minutes — is where you engage with anything you need to address: a piece of developmental feedback that came up since your last meeting, the organizational context the person needs to understand, or an upcoming decision where their input matters. The final five to ten minutes are forward-looking:

what is the person working toward in the next week, what might get in their way, and what support do they need from you specifically before you meet again?

Development conversations deserve their own protected space within the one-on-one rhythm. A useful practice is to designate one one-on-one per month — perhaps the last one of the month, or whichever cadence fits your schedule — as a development-focused session. This is not a performance evaluation session; it is a growth conversation. You review together the progress on the individual development plan, discuss skills the person is actively working on, explore what is opening up or closing off in their longer-term career trajectory, and identify new challenges or stretch assignments that could accelerate their growth over the next quarter. By separating these conversations from the operational rhythm, you prevent development from being perpetually crowded out by the urgency of immediate work, which, if left to chance, is exactly what happens.

Several structural commitments make one-on-ones materially more effective than the default. First and most importantly, protect them. Canceling a one-on-one to attend an unnecessary meeting sends a signal about priorities that is difficult to walk back. Second, take notes during the conversation and review them before the next session. Nothing undermines trust faster than consistently forgetting what was discussed — including commitments you made to the other person. Third, follow through on every commitment, every time. If you said you would send a resource, connect them with a stakeholder, review their draft, or ask a question on their behalf, do it before your next meeting. Fourth, when possible, rotate the format — walking one-on-ones, coffee off-site, or a change of environment can unlock candor that does not emerge in a conference room or on a video call.

Diagram 7.5: One-on-One Meeting Structure

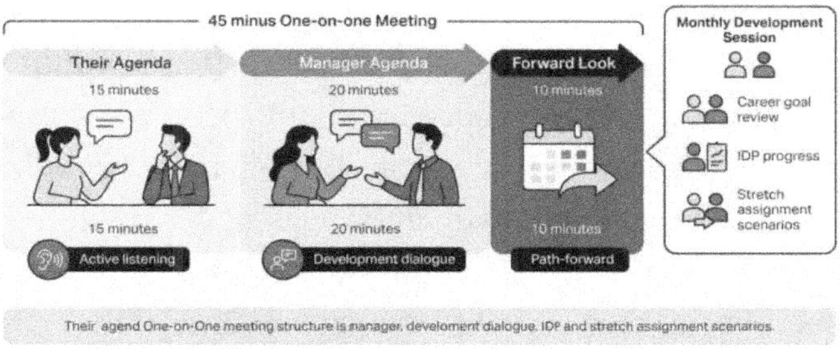

8.7 Building Individual Development Plans That People Believe In

The individual development plan has an unfortunate reputation in most organizations, and it is largely deserved. People associate IDPs with bureaucratic compliance theater — a template that gets filled out during the annual performance cycle, reviewed once by an HR system, filed in a digital folder where no one will see it again, and referenced only when someone needs documentation for a promotion case or a performance improvement process. That version of the IDP is not just useless — it is actively damaging because it associates formal development planning with organizational inauthenticity. When you introduce an IDP to your team members, you may have to overcome exactly this association before you can build something genuinely useful.

The meaningful IDP — the one that actually produces development — is built around four generative questions rather than a compliance form. First: where does this person genuinely want their career to go? Not where you think they should go, not where there happens to be an organizational opening, not the default trajectory that most people in similar roles follow. Where do they actually want to take their professional life? This question often requires real investment to answer, because many people — especially those early in their careers or coming from organizational cultures where this conversation was never invited — have not been given explicit permission to think about their own trajectory in the context of their work. They may need several conversations before they have a clear enough sense of their own aspirations to articulate them. That is fine. The clarity you build together in those conversations is development in itself.

The second question is: what capabilities does getting there require, and which of those capabilities are currently gaps relative to where this person is now? This is where AI-powered skills gap analysis tools begin to add meaningful value — they can help you surface the competency landscape for target roles with specificity and currency that a manager's intuition alone cannot match. The third question is: what specific development activities — not just training courses but also stretch assignments, mentoring relationships, cross-functional projects, conference presentations, community leadership, and deliberate practice — will close those gaps over a realistic time horizon? And the fourth question is: how will we measure progress, and when will we check in on it?

The most important behavioral principle in IDP construction is co-creation. The IDP that you write for someone, based on your own assessment of their gaps and your own sense of their potential, and hand to them to sign off on, is not a development tool — it is an assignment. The IDP that the person drafts themselves, in response to a genuine exploration of their own goals and gaps, with your guidance, challenge, and support in shaping it, is a commitment they own. That ownership difference is not trivial. People pursue with real energy what they experience as theirs. They comply halfheartedly with what they experience as imposed, no matter how well-intentioned the imposition.

Stretch assignments are the highest-return development vehicle available to most managers, and they require no training budget, no external vendor, and no HR approval. Giving a team member ownership of a cross-functional project that exposes them to stakeholders they would not otherwise engage with. Asking an emerging leader to represent the team in an executive briefing on a topic they know well. Having someone with leadership potential run a team process-improvement initiative with direct accountability for the outcome, and pairing a less-experienced team member with a more experienced one in a structured mentoring arrangement focused on a specific skill domain. These are development investments that simultaneously advance real work and grow real people. The key is matching the stretch to the person's current capability and their growth edge — the zone where the challenge is demanding enough to require new capability, but not so far beyond current capacity that failure is the most likely outcome.

Diagram 7.6: Individual Development Plan Co-Creation Framework

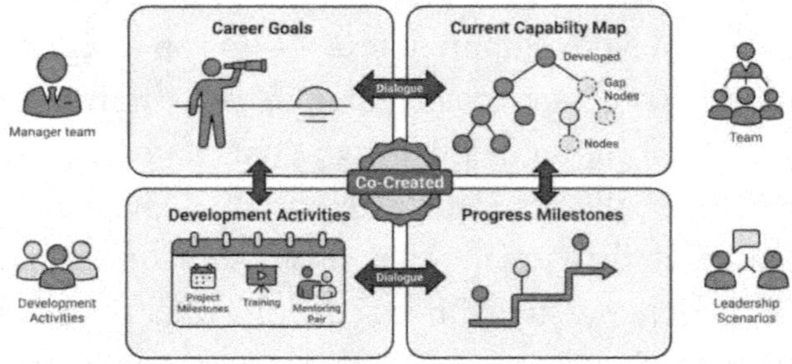

Collaborative framework facilitates aligned growth through continuous dialogue between manager and team member.

8.8 AI-Driven Skills Gap Analysis and Personalized Learning

Until recently, organizational skills gap analysis was a laborious, expensive, and frequently inaccurate exercise. Managers conducted informal observations of the work happening around them and formed intuitive assessments of who had what. HR ran periodic competency assessments based on frameworks that were often outdated relative to the actual skill demands of current and emerging roles. Learning and development teams made educated guesses about what training the organization needed based on manager surveys and attrition interviews. The results were often disconnected from actual workflow requirements, arrived months after the gaps were most acute, and failed to account for the diversity of individual learning preferences, paces, and career trajectories that characterize any real team. AI-powered skills intelligence tools are fundamentally changing this equation, and the change is accelerating quickly enough that managers who are not aware of these capabilities are making decisions with significantly less information than their peers.

Modern AI-enabled learning platforms — tools like Degreed, LinkedIn Learning's AI-powered features, Coursera for Business, Workday Learning, and Cornerstone's AI suite — can analyze a team member's role requirements, current verified skill profile, stated career goals, past learning behavior and completion patterns, and performance data to surface learning recommendations that are specific, timely, and genuinely relevant to the person's situation. These platforms increasingly ingest signals from outside formal learning as well: projects completed and their stated competency requirements, certifications earned and their recency, skills endorsed or validated by peers, and language patterns from roles the person has expressed interest in or been assessed against. The result is a skills intelligence picture that is more current, more granular, and more individually tailored than any periodic human-conducted assessment could realistically produce.

For managers, the most immediately actionable AI capability in the skills domain is team-level skills gap detection — the ability to identify mismatches between the team's current aggregate skill profile and the competency requirements implied by the team's upcoming roadmap, strategy, or organizational direction. Rather than waiting for a project to struggle or fail before discovering that a critical capability is missing, AI tools that have been configured with your team's current skill data and your planned work can surface those mismatches proactively — weeks or months before the gap becomes a performance problem.

This is what operational clarity in the AI era actually looks like: knowing what you do not have before you need it, and having enough lead time to address the gap through training, hiring, contracting, or strategic de-scoping.

A few cautions are warranted here, and they deserve serious weight. AI learning and skills recommendations are only as good as the data on which they are trained and the historical patterns from which they learn. If the high performers in your organization historically share certain demographic characteristics — and in many industries and organizations, they do — an AI skills assessment system trained on that data will systematically learn to value those characteristics as predictors of potential, which may encode and perpetuate the organizational biases that already existed. More subtly, AI systems trained on historical career pathways tend to recommend traditional development routes rather than novel ones, inadvertently narrowing the development imagination of both managers and team members. Treat AI skills assessments as one important input among several. Combine them with your own direct observation of the person's work, the person's own self-assessment of their strengths and gaps, and candid input from peers and collaborators who see different dimensions of their performance. The manager remains the integrator of signals — not the passive consumer of an algorithm's recommendation.

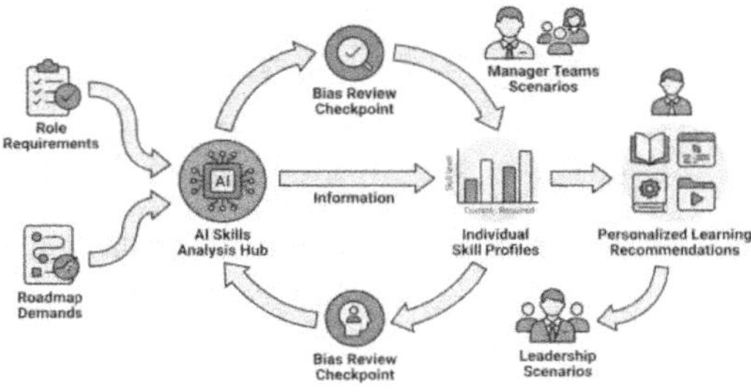

This continuous cycle enables accurate skills intelligence and personalized development with integral bias oversight.

8.9 Mentoring, Coaching, and Sponsoring — Knowing the Difference

One of the most pervasive errors in talent development practice is treating coaching, mentoring, and sponsoring as interchangeable terms for a general category of developmental support. They are not. They serve different purposes, operate through different mechanisms, produce different outcomes, and are appropriate in different circumstances and at different career stages. Conflating them does not just reduce their effectiveness — it can actively undermine the development relationship by applying the wrong instrument to the situation. The manager who defaults to advice giving when a coaching conversation is what is needed robs the person of the chance to develop their own judgment and problem-solving independence. The manager who coaches endlessly without ever exercising organizational power on someone's behalf fails to honor the positional influence they have fully. The manager who mentors when sponsorship is what would actually move the person's career forward provides comfort but not traction.

204

Coaching, as explored throughout this chapter, is a structured, question-driven conversation designed to develop the other person's own thinking, awareness, and capability. The coach does not transfer their own experience or make the decision — they create the conditions for the coachee to access their own wisdom, clarify their own goals, and construct their own path forward. Coaching is the right mode when the person can work through the challenge independently and needs a structured thinking partner to help them do so. It is also the right mode when the developmental priority is to grow the person's problem-solving independence rather than to solve the immediate problem most efficiently. If every time a team member comes to you with a problem, you tell them what to do, you have not developed a team member — you have developed a dependency.

Mentoring is advisory and experiential. It draws directly on your accumulated experience and hard-won perspective to help someone navigate territory you have already crossed. A mentor shares how they navigated a similar organizational politics challenge, what they wish they had known at a particular career stage, which professional relationships proved most valuable and how they were cultivated, and what risks they would take if they were starting over with what they know now. Mentoring is the right mode when the person is genuinely navigating unfamiliar territory, and your direct experience is relevant and applicable. The risk to manage actively is that mentoring can slide into prescriptiveness — telling the person exactly what to do rather than sharing your experience as one reference point among many, which short-circuits their own learning and creates a dependency on your approval before they act.

Sponsoring is the most underutilized of the three, particularly for team members who lack organizational visibility, access, or social capital. When you sponsor someone, you actively leverage your own credibility, connections, and positional power to advance their career in ways they cannot on their own. You recommend them for high-profile projects. You say their name when an opportunity comes up in a room they are not in. You invite them to present their own work to senior leadership rather than presenting it on their behalf. You make a deliberate, active introduction to someone in your network whose relationship would benefit their career. You argue for their promotion when their name comes up in calibration discussions that they are not part of. For team members who are highly capable but underrecognized — which disproportionately includes people from historically underrepresented groups in leadership — sponsorship is not a nice-to-have. It is the mechanism through which systemic access inequality gets corrected at the individual level. Ask yourself, regularly and honestly: whose name am I saying when I am not in the room with them?

8.10   Manager's Checklist — Coaching and Developing Your Team

Use this checklist as a weekly and monthly operational reference. These behaviors are not aspirational ideals appropriate only for senior or experienced managers. They are foundational operating practices that distinguish managers who actually develop talent from those who manage tasks while talent development happens by accident—or not at all.

- Hold all one-on-ones on a consistent, protected schedule — cancel them only for genuine emergencies and reschedule within the same week
- Begin each one-on-one by asking what the team member wants to cover before adding your own agenda items
- Designate one one-on-one per month as a development-focused session — come prepared with questions about career goals and growth

- Use the GROW model structure in coaching conversations — ask before telling, and ask more than you think you need to
- Use SBI structure whenever you deliver corrective or developmental feedback — Situation, Behavior, Impact, followed by Inquiry
- Deliver positive feedback specifically, soon after the observed behavior, and often publicly
- Deliver developmental feedback specifically, privately, calmly, and close in time to the observed behavior
- Ask for feedback from your direct reports at least once per month — respond to what you receive with visible action or acknowledgment
- Ensure every team member has an active IDP that was co-created through a genuine conversation about their goals
- Review each IDP quarterly — update development activities as the person's goals, capabilities, or organizational context change
- Identify at least one stretch assignment opportunity for each team member per quarter
- Consciously distinguish your mode before each development conversation — are you coaching, mentoring, or sponsoring?
- Ask yourself monthly, for each team member: whose name am I saying when I am not in the room with them?
- Use AI-powered learning and skills tools as one input among several — combine with direct observation and the person's own self-assessment

- Monitor engagement and motivation signals across your team — IDP participation, one-on-one quality, feedback receptiveness — as leading indicators of retention risk

8.11   The Takeaway

The manager who consistently and deliberately develops talent is practicing the highest form of organizational leverage available to anyone in a leadership role. Every hour invested in a coaching conversation that expands a team member's self-awareness yields a compounding return. Every piece of feedback that lands with precision and care, redirecting a behavior that would otherwise have continued unchecked, is an investment with a measurable business consequence. Every individual development plan built with genuine co-creation and reviewed with genuine accountability is an investment in retention, capability, and team resilience that no amount of reactive talent acquisition can replicate.

The coaching mindset is not a personality trait you either have or do not have. It is not a gift that some managers are born with, and others are permanently without. It is a set of practices, frameworks, and habits of attention that any manager can develop through deliberate effort, honest self-reflection, and the willingness to be uncomfortable while learning. Start your next one-on-one with the GROW model, where someone brings you a problem they could solve themselves. Use SBI once this week, completely, and notice what happens differently in the conversation. Designate your next end-of-month one-on-one as a development session and walk in with questions about the person's career aspirations rather than a project update. These are not large behavioral changes. They are small, deliberate shifts that, repeated consistently over weeks and months, produce fundamentally different outcomes for the people you lead.

AI will continue to change the surface features of the coaching and development landscape — the tools for skills gap analysis, the mechanisms for feedback collection and aggregation, the platforms through which personalized learning is delivered and tracked. What it will not change is the fundamental human reality: people grow fastest when they feel genuinely seen by someone who has invested in knowing them, challenged in ways that expand their capability at the edge of their current reach, and supported by a manager who is consistently and credibly in their corner. That investment is irreducibly human. No algorithm can replicate the impact of a manager who remembers what someone said they wanted to achieve three months ago and then actually makes space for that conversation — again and again and again.

9 Managing Conflict and Navigating Difficult Conversations

9.1 A Scenario Worth Sitting With

Six weeks into his new role as an operations manager, Derek inherited a team that had been under significant strain. Two senior analysts — Yolanda and Marcus — were barely speaking to each other. Their disagreement had originated three months earlier over a process redesign initiative, in which each had staked out opposing positions before senior leadership. The working silence between them had begun to fracture the broader team: people were picking sides, information was being withheld across an informal boundary that had no place on an organizational chart, and the friction was starting to affect delivery timelines for commitments the team had made publicly. Derek knew he needed to address it. Every management instinct he had told him this situation would not resolve itself. But every time he imagined the conversation, he felt a wave of preemptive avoidance: What if it got worse? What if one of them escalated to HR and the situation became formally documented? What if addressing it directly destroyed the fragile functional peace that currently holds? He told himself he needed more information. More time. A better moment.

Derek's hesitation is entirely understandable and extraordinarily common. New managers are almost universally undertrained in conflict resolution. The organizational systems around them frequently incentivize avoidance — conflict is noisy, time-consuming, emotionally taxing, and unpredictable in outcome, while the path of least resistance is to wait, hope the friction subsides on its own, and direct attention toward work that feels more manageable. The problem is that interpersonal conflict in professional environments rarely resolves itself. What begins as a contained friction point between two individuals expands reliably over time. It absorbs team bandwidth, degrades psychological safety across the entire group, frequently triggers the voluntary departure of one or both parties, and often generates collateral disengagement among team members who are not directly involved but who absorb the tension daily. By the time a conflict reaches the point where a manager feels they can no longer avoid it, it has usually already cost far more in lost productivity, eroded relationships, and diminished team morale than early intervention would have.

This chapter gives you the frameworks, language, and mental models to address conflict proactively, to conduct difficult conversations with both skill and composure, and to maintain team cohesion when interpersonal tension, organizational change, or AI-related disruption threatens it. It covers the full spectrum of conflict management: prevention through culture and structure, early detection before issues escalate, active intervention when they do, and the specific and increasingly common challenge of leading teams through AI-driven change that generates its own category of professional anxiety and interpersonal friction. Difficult conversations, handled well, are not team breaking events. Handled consistently well, over time, they become the mechanism through which trust is built, norms are clarified, and teams develop the resilience to navigate whatever comes next.

Diagram 8.1: Conflict Escalation Spectrum

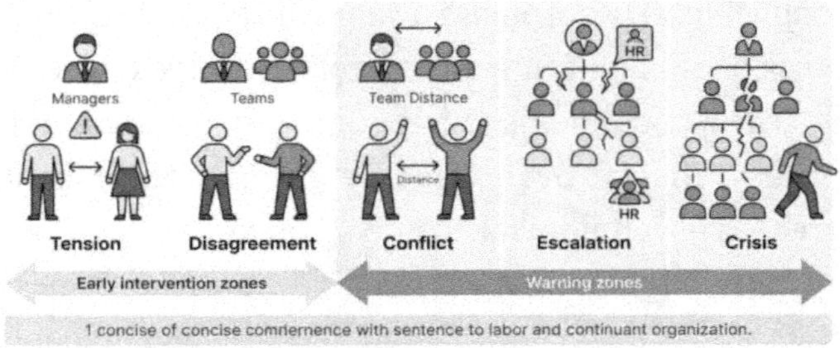

9.2 Why This Matters

Unresolved conflict is among the most expensive silent costs in any organization, and it is silent precisely because the damage accumulates beneath the surface of what shows up in productivity metrics, project status reports, and performance reviews. The direct costs — time spent in unproductive interactions, work duplicated because communication has broken down across a conflict boundary, decisions delayed because conflicted parties avoid the collaboration required to make them — are real, measurable, and compounding. The indirect costs are harder to quantify and substantially larger. Psychological safety, the organizational condition that Google's landmark Project Aristotle research identified as the single most important factor in high-performing teams, is nearly impossible to maintain in an environment of unaddressed interpersonal conflict. When team members perceive that speaking up carries the risk of bringing them into a conflict dynamic — being aligned with one side or the other, or being seen as the person who raised the issue — they stop speaking up. Ideas go unspoken. Problems go unreported. Risk goes unacknowledged. The consequence is not just interpersonal discomfort; it is strategic blindness.

For the new manager specifically, conflict avoidance carries a compounding organizational risk that extends well beyond the immediate situation. Early in your tenure, you are establishing the behavioral norms of your team through your own actions and responses. How you respond to the first visible interpersonal conflict on your watch sends a signal that echoes through the entire team about what will and will not be tolerated, what the manager considers their responsibility to address, and what the consequences of bringing a concern to this leader are likely to be. A manager who lets conflict fester is signaling — unintentionally, but unmistakably — that organizational peace is more valuable than organizational health, that whoever escalates first will be labeled the troublemaker, and that the safe behavior is to endure rather than address. A manager who addresses conflict directly, fairly, and without delay signals that this is a team where hard things are handled with maturity and accountability, where both parties are heard, and where the manager's job is to create conditions for people to work effectively rather than to keep a lid on discomfort.

AI adoption introduces a significant, often underrecognized, new category of workplace conflict that managers of AI-era teams must understand and navigate. When an organization begins integrating AI tools into team workflows, it almost always triggers a set of individual anxieties and interpersonal tensions that are genuinely novel in character. Team members have concerns about job security that may be well-founded or exaggerated, but are rarely openly discussed. They worry about the valuation of their existing skills in a landscape where AI can perform some of those skills more quickly, and they may respond to that worry by either overclaiming their AI capabilities or digging in defensively against adoption. They observe that AI adoption benefits are distributed unevenly across the team — some roles change dramatically, others barely at all — and they develop concerns about fairness that surface indirectly, as resentment rather than explicit conversation. The interpersonal friction these dynamics generate often appears, on the surface, to have nothing to do with AI at all. Recognizing the AI-adjacent conflict signal beneath what appears to be a routine interpersonal issue is a skill that distinguishes managers who can lead effectively through digital transformation from those who are managing the symptoms while the underlying dynamics worsen.

9.3 Understanding Sources and Types of Workplace Conflict

Not all workplace conflict looks the same, and treating different types of conflict with the same intervention approach is a management error that frequently makes things worse. There are at least four meaningfully distinct sources of workplace conflict, each of which calls for a different diagnostic lens, a different conversation style, and a different resolution approach. Understanding which type of conflict you are dealing with before you intervene is the difference between a focused, productive intervention and an unfocused conversation that generates more heat than clarity.

Task conflict arises when team members disagree about the content, approach, methodology, or quality standards of work itself. This is the most productive form of conflict when managed well — teams that engage in substantive, evidence-based debate about ideas and approaches make better decisions, catch more errors, and produce higher-quality outcomes than teams that achieve false consensus through social pressure or deference. The manager's role with task conflict is not to eliminate it but to structure it productively: establish norms for how substantive disagreements get raised and resolved, ensure that all voices — including those of less organizationally powerful team members — are heard in the debate, clarify decision rights so that people know how and by whom a final call will be made, and distinguish between debates where the goal is the best idea (which benefits from ongoing challenge) and decisions that need to be made and executed even if not everyone agrees.

Process conflict concerns how work gets done — who is responsible for what, which procedures govern a given workflow or decision, how resources are allocated across competing needs, and what accountability structures are in place when something does not go as planned. Process conflict is chronically underestimated as a source of interpersonal tension in organizations. Unclear roles and ambiguous decision rights generate enormous friction that manifests as personality conflict but has a structural root.

When two experienced professionals are both convinced they own a particular decision — because the organizational design left that ownership unclear — the result is either a standoff, a covert power struggle that consumes organizational energy for months, or one party's capitulation that breeds ongoing resentment. The manager's intervention for process conflict is primarily structural: clarify roles explicitly, define decision rights in writing, and establish transparent processes that remove the ambiguity that made the conflict predictable in the first place.

Relationship conflict is personal and almost always the most difficult to address, partly because it is the most visible, partly because it is the most uncomfortable to engage, and partly because the parties involved often have well-developed narratives about each other that feel like facts rather than interpretations. When a series of interactions has generated a narrative of mistrust — when one person genuinely believes the other is unethical, incompetent, or personally hostile — the conflict has moved beyond tasks and processes into a relational dynamic that requires a different level of managerial engagement. The critical insight about relationship conflict is that, as the manager, you are not required to make people like each other. You are required to make them functionally collaborative professionals who can do the work that serves the organization and the team. That is a more achievable goal, and framing it clearly — to yourself and to the parties involved — often makes the intervention significantly more productive.

Value conflict — disagreements rooted in fundamentally different beliefs, ethics, priorities, or views of what is right — is the rarest but most intractable form of workplace conflict. When a team member objects to a policy on genuine ethical grounds, or when two people's fundamental beliefs about fairness, respect, or professional integrity put them in sustained opposition, no facilitation technique built on finding common ground is likely to resolve the disagreement at its root. The manager must make a judgment about what is required of all team members as professionals, regardless of personal values, the organization's stated values, and whether a persistent value misalignment is ultimately incompatible with continued productive membership on the team. These are not easy judgments, and they sometimes require formal HR engagement and legal guidance from the organization.

Diagram 8.2: Four Sources of Workplace Conflict

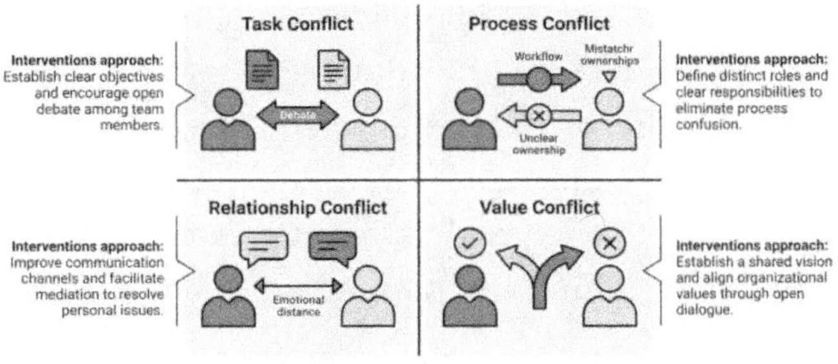

Understanding these four sources enables effective conflict management and stronger leadership.

9.4 The Conflict Resolution Framework — Prevention to Intervention

The most effective conflict management strategy is to invest heavily in creating conditions in which destructive conflict is structurally unlikely to emerge in the first place. This prevention layer is not glamorous — it involves decisions about team design, communication norms, and process clarity that rarely generate visible organizational credit — but it is the layer with the highest return on managerial investment. Teams that have done the prevention work handle the conflicts that do arise more quickly, with less collateral damage, and with stronger relationships intact afterward.

The first prevention practice is establishing team norms explicitly and collaboratively. Teams that have documented, discussed, and periodically revisited their shared expectations for how they operate — how disagreements are raised, how decisions get made once the debate is complete, what it looks and sounds like to be a respectful and accountable team member in this specific group — have a materially lower incidence of destructive conflict than teams that operate entirely by implicit assumption. If you have not yet led a team charter or working agreement exercise, do it now. If you already have one, when did you last look at it? Norms that are created and never revisited quickly drift away from actual behavior, and a gap between stated and actual norms is itself a source of tension.

The second prevention practice is managing the structural conditions that breed conflict before they generate interpersonal friction. Most workplace conflict that presents as personal incompatibility have a structural root that preceded it and created the conditions for it. Role ambiguity, uneven workload distribution, competing performance incentives that reward individual success at the expense of collective outcomes, and resource scarcity that forces implicit competition — these are organizational design failures that inevitably produce interpersonal friction. Before attributing conflict to personality or chemistry, conduct a structured analysis: what organizational condition makes this conflict predictable? Frequently, the answer reveals that the solution is structural rather than interpersonal, and that the interpersonal intervention alone, without addressing the structural root, will produce a temporary improvement followed by a return of the same dynamics.

The third prevention practice is maintaining genuine visibility into team dynamics through consistent one-on-one conversations that are candid enough to surface early signals. Conflict almost always produces observable early indicators before it becomes visible as an explicit interpersonal problem. When you hear in a one-on-one that someone feels consistently dismissed in team meetings, or that two specific people have been avoiding a collaborative task they would normally share, or that someone is carrying a frustration about a decision they experienced as unfair but never raised directly, these are conflict signals still in the prevention zone. Acting on them thoughtfully, curiously, and without accusation or over-reaction is the difference between catching a spark and fighting a fire.

When conflict has moved beyond prevention and requires active intervention, the process framework that produces the best outcomes has three distinct phases that must be executed in sequence. The first phase is individual conversations with each party separately, before any joint session. Before you bring people into the same room, you need to understand each person's perspective in their own words: what they believe happened, how it has affected them, what they want from a resolution, and what they are willing to contribute to getting there. This investment serves two purposes: it gives you the intelligence you need to facilitate effectively, and it signals to each person that they will be genuinely heard rather than managed. The second phase is a structured joint conversation, if and when the parties are ready to engage in productive dialogue. Not all conflicts benefit from bringing the parties together — the judgment about readiness requires knowing each party's emotional state, the nature and history of the conflict, and whether there is enough shared goodwill to make joint dialogue productive rather than escalatory. The third phase is follow-through: explicitly checking in with both parties over the following weeks, acknowledging when the relationship is visibly improving, and addressing any backsliding before it compounds.

Diagram 8.3: Conflict Intervention Process

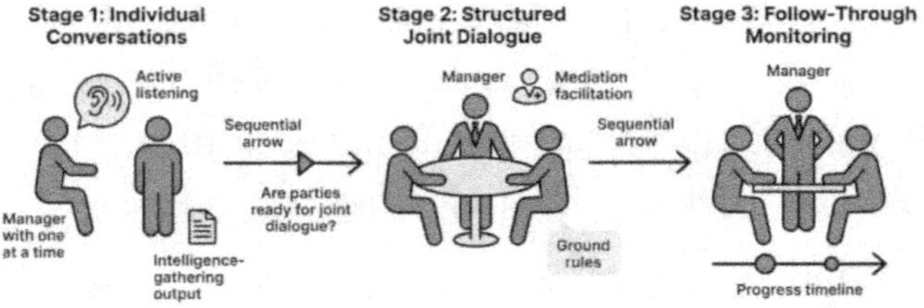

The process friendy with one entinoudian ronversations and a for working together of conflict resolution.

9.5 Having Difficult Conversations —
Performance, Terminations, Sensitive
Topics

Difficult conversations are not all the same in nature, stakes, or required skill set. But they share a common underlying architecture that, when applied consistently, makes them more productive than avoidance and less damaging than poorly executed immediacy. The three elements of that architecture are preparation, groundedness, and follow-through. Preparation means knowing with precision what you need to communicate, why it matters organizationally and personally, what outcome you are seeking from this specific conversation, and how you will respond to the predictable emotional and logical responses the other person is likely to have. Groundedness means entering the conversation from a place of calm, factual clarity rather than anxiety or accumulated frustration, which requires that you have done the emotional preparation — including processing your own feelings about the situation — as well as the informational preparation. Follow-through means honoring whatever commitments you made in the conversation and maintaining the professional relationship through the uncomfortable aftermath that almost always follows a difficult exchange.

Performance conversations are among the most frequent difficult conversations a manager faces across a career, and the most common error is waiting too long to have them. Managers delay for understandable reasons: they hope the situation will improve without intervention, they do not want to be the person who delivers bad news, they are not certain enough in their assessment to feel they can defend it under challenge, or they genuinely like the person and dread the emotional weight of telling them their performance is not meeting expectations. But every week of delay makes the conversation harder, the gap between expectation and reality larger, the organizational consequences more severe, and the eventual intervention more formal and more consequential for everyone involved. The performance conversation that happens at week three of a problem is a brief, supportive, corrective exchange. The same conversation in week fifteen is part of a formal performance improvement process.

The performance conversation formula that works is simple enough to be implemented immediately and specific enough to produce meaningful outcomes. State the performance expectation clearly — not as a reminder that the expectation exists, but as a precise description of what the behavior or output looks like when it meets the standard. Describe the gap between that expectation and what you have observed, using SBI structure to ground the description in specific, observable, verifiable reality rather than impression or interpretation. Express your genuine confidence that the person can close the gap — if you do not have that confidence, you are having a different conversation than a development-oriented performance discussion. Establish a clear, measurable, time-bound picture of what improved performance looks like, and schedule a follow-up check-in that signals your investment in the resolution rather than your preparation for a formal process.

Termination conversations are categorically different from performance conversations and deserve separate preparation and a fundamentally different orientation. The primary principle is that the termination conversation is not a negotiation, not a final warning, and not a venue for additional performance feedback. The decision to separate the person has already been made — through a process that likely included HR, a legal review, documentation of performance or conduct issues, and any organizational approval your company requires. The termination conversation is a notification, delivered with respect and as much dignity as the circumstances allow. Keep it clear and brief. Do not over-explain, apologize at length, or narrate the decision process in detail — this creates confusion about whether the decision is actually final and invites negotiation. Make the practical logistics unambiguous: what happens with compensation, benefits, equipment return, and system access. Provide a specific contact for follow-up questions. The manner in which someone is separated from your team is observed by every person who remains. It shapes what your team believes about whether they will be treated with dignity if they are ever in a similar position.

Sensitive conversations — those involving mental health concerns affecting work performance, personal circumstances creating professional impact, perceived bias or discrimination, religious or cultural accommodation requests, or serious interpersonal complaints — require a distinctly different orientation than either performance or termination conversations. The primary stance is curious and supportive rather than evaluative or managerial. Your first job is to listen carefully, ask open-ended questions that create space for the person to share what they are experiencing at whatever level of detail they choose, and resist the impulse to jump to solutions, reassurances, or explanations before you have genuinely heard the situation. Know the boundary of your role: you are a manager, not a therapist, a lawyer, or an HR professional. There are circumstances — and more of them than most new managers realize — where the most valuable thing you can do is acknowledge what you have heard and connect the person with an appropriate professional resource, rather than attempting to solve the situation yourself.

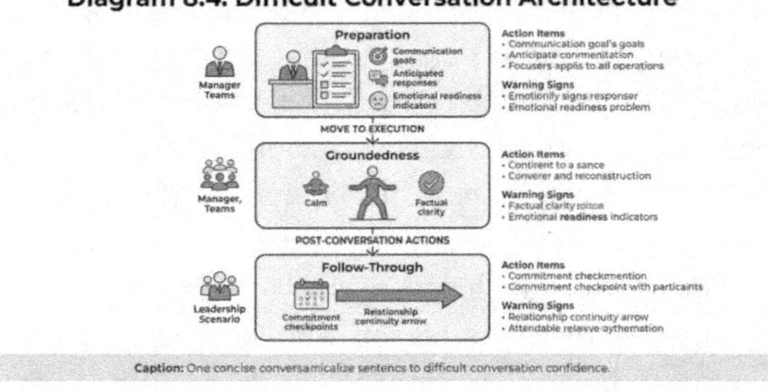

Caption: One concise conversamicalize sentencs to difficult conversation confidence.

9.6 Managing Up — Navigating Disagreements with Your Own Manager

The difficult conversation that many new managers find most daunting is not the one they need to have with a direct report but the one they need to have with their own manager. Managing up — advocating for your team when their needs are being underserved, pushing back on a decision you believe will produce poor outcomes, raising a concern about an organizational direction you think is misaligned with the stated strategy, or requesting resources that your team genuinely needs to do its work effectively — is a critical management skill that most management development programs barely address. The instinctive responses to disagreement with a superior are either capitulation, which over time erodes your own credibility and your team's trust in your advocacy, or avoidance, which produces the same slow erosion through a different mechanism. Neither serves the manager, the team, or the organization.

The most effective managing up practice begins with a disciplined exercise in perspective-taking before you say a word. Before you raise a concern or advocate for a different decision, invest real cognitive effort in understanding your manager's own situation: what are the organizational pressures and constraints shaping their decisions, what do they care about most deeply and what are they being held accountable for by their own manager, what would have to be true for the decision you are disagreeing with to be reasonable from where they sit in the organization? This exercise is not about abandoning your own position or pre-capitulating to authority. It is about entering the conversation with enough genuine understanding of their perspective to engage their actual reasoning rather than the version of their reasoning that you have imagined and are prepared to argue against.

The language of effective managing up is problem-framing and solution offering, not complaint making. 'I want to share a concern about our capacity to deliver on the expanded scope — I've been looking at the team's current commitments, and I think we need to either adjust the timeline or identify what we can de-prioritize' is a managing up statement that frames a real problem and signals your readiness to participate in solving it. 'This is too much, the team is already overwhelmed, and this is an unreasonable ask' is a complaint that invites defensive justification rather than collaborative problem-solving. The difference in organizational impact between these two versions of the same underlying concern is substantial. Bring data where you have it. Bring two or three concrete alternatives rather than a single objection. Acknowledge explicitly what you understand about their constraints before making your ask. And choose the moment carefully — a hallway conversation during a high-pressure deadline period is not the place to raise a concern that deserves serious, focused engagement.

There is a distinct category of managing up situation that requires a different level of seriousness and preparation: the circumstance where you believe that a decision or direction is genuinely wrong — not just suboptimal from a business perspective, but ethically problematic, potentially harmful to people inside or outside the organization, or possibly in violation of legal or regulatory requirements. These situations are uncomfortable and genuinely consequential, but they represent exactly the professional obligation that comes with any management role. Document your concern clearly and factually before raising it. Understand your organization's formal escalation channels, including compliance hotlines, ethics reporting mechanisms, legal department contacts, and HR processes. Consult a trusted mentor or experienced colleague before taking formal action if you can do so without compromising the confidentiality of the situation. The manager who raises a legitimate ethical concern through appropriate channels, with documentation and good faith intent, is fulfilling the organizational responsibility that comes with the authority of their role. The manager who remains silent to avoid personal discomfort has made a different choice about what that authority obligates them to.

9.7 Handling Resistance to AI Adoption and Technology Change

AI adoption is fundamentally and inescapably a human change before it is a technology change. When your organization rolls out a new AI-powered tool, workflow, or process, you are not asking your team to learn a new piece of software the way they once learned to use a new project management platform or a new customer relationship system. You are asking them to reconsider their relationship with their own professional expertise, to reassess what their skills are worth in a world where AI can replicate some of those skills faster and at lower cost, and to trust an opaque, probabilistic system with tasks they have historically owned and controlled and taken professional pride in. The resistance that follows these asks is not irrational, not obstructionist, and not unique to a certain kind of employee. It is the predictable, largely appropriate human response to a genuinely disorienting change in professional self-concept and in organizational power dynamics.

The first responsibility of the manager in an AI adoption context is to resist the temptation to characterize all resistance as a single phenomenon with a single cause and a single solution. Resistance is not monolithic. One team member may be resistant because they have direct, specific experience with an AI-generated error that produced a real consequence for work they were accountable for. Their resistance reflects reasonable professional skepticism rather than technophobia. Another may be resistant because the new AI workflow creates evaluation pressure that did not previously exist — their work is now being compared against an AI baseline, and the comparison feels unfair. A third may be resistant because the adoption timeline is genuinely unrealistic relative to their current workload, and compliance would require them to cut corners on work quality that they care deeply about. These are three completely different problems, rooted in distinct concerns, requiring three distinct conversations. Treating all resistance as an attitude problem to be corrected rather than information to be understood is a management error that damages trust and slows adoption more reliably than any technical implementation challenge.

The framework for leading through AI adoption resistance has four sequential steps, each building on the previous one. First, validate the concern — not the conclusion the person has drawn, but the underlying professional worry that is generating the resistance. 'I hear that you're concerned about accuracy in edge cases — let's look at the specific examples together and understand where the tool performs well and where it needs human oversight' is a validating response that opens a productive dialogue. 'Everyone else is adapting fine, I need you to get on board' is a dismissive response that closes it and adds resentment to the existing resistance. Second, create structured, low-stakes experimentation opportunities. People who are resistant to AI adoption become measurably less resistant when they are allowed to experiment with a tool on a non-critical task in a context where their evaluation is not immediately tied to the experiment's outcome. Resistance often reflects anxiety about the unknown, and direct experience with the tool at low stakes is the most reliable antidote to that anxiety.

Third, surface and celebrate early wins from the adoption — not as a mechanism to pressure holdouts through social comparison, but to shift the conversation from abstract anxiety about what might happen to concrete evidence about what is actually happening. When one team member reports that the AI tool saved them four hours on a recurring report that they have always found tedious, and that reclaimed time is now going toward more interesting and impactful work, that story is more persuasive than any adoption mandate. Fourth, and perhaps most importantly, address the legitimate concern that some aspects of team roles may genuinely change as a result of AI adoption. The reassurance that 'this tool is just to help you, it won't change your job' is often false, and team members who are perceptive enough to have noticed that it is false will trust you less for having said it.

Honesty about workflow evolution — 'this tool will change how you spend your time, and we're going to work together to understand what that means for how your role develops' — is more trust-building than a comforting narrative that does not survive contact with reality.

9.8 De-escalation Techniques and Mediation Skills

When conflict has escalated to the point where emotions are elevated and rational, collaborative problem-solving has broken down — when people are defending positions rather than exploring interests, when the conversation is generating heat rather than light — the manager's first job is de-escalation. Not resolution, not mediation, not solution finding. De-escalation. Attempting to resolve a substantive problem during an emotional activation does not produce resolution. It produces more activation, more entrenched positions, and a set of commitments — made under emotional pressure — that neither party will honor once they have had time to reflect. The intervention that produces durable outcomes creates sufficient emotional space for the parties to access their rational, problem-solving capacities before attempting to use those capacities.

The most reliable de-escalation technique available to a manager is genuine active listening — not performed listening in which you visibly nod while formulating your response, but the genuine absorption of what the other person is communicating, including both the informational content and the emotional signal beneath it, followed by a paraphrase that demonstrates real comprehension. 'What I'm hearing is that you felt your contribution to the design process was invisible to the team, and that the way the decision was announced afterward compounded that feeling by making it look like the decision was made without you. Is that an accurate understanding?' This technique works because it redirects emotional energy from continued expression into the experience of being understood. When people genuinely feel heard — not agreed with, but heard — the urgency to escalate decreases sharply, and the cognitive bandwidth for problem-solving increases.

A second technique that reliably supports de-escalation is asking process questions rather than content questions. Content questions — 'What actually happened?' or 'Who was responsible for that decision?' — invite the parties to relitigate the events that generated the conflict, which almost always produces competing narratives, defensive justifications, and an escalation of the original emotional charge. Process questions redirect attention from the past, where the argument lives, to the future, where resolution is possible: 'What would you need to see happen to feel that this situation had been addressed fairly?' or 'What would a workable working relationship between you and Marcus look like, and what would have to change for that to be possible?' Process questions also surface the actual interests — the underlying needs and concerns — beneath the stated positions, which is almost always where the space for resolution exists.

When you are facilitating a direct conversation between two conflicted parties — rather than working with each person individually — the mediation orientation is neutrality without passivity. Neutrality means you are not there to determine who is right, validate one party's narrative, or advocate for a particular outcome. Passivity means you sit back and let the conversation go wherever the emotional dynamics of the moment take it. Neither is effective. Active neutrality means you are present and engaged as a facilitator — establishing and enforcing ground rules explicitly, redirecting when the conversation moves toward personal attacks rather than specific behaviors, summarizing areas of agreement and identifying remaining points of disagreement, and actively keeping the conversation oriented toward the forward-looking question of what a workable relationship and a functional collaboration look like from this point. It also means being willing to pause or end a joint session if the temperature is rising rather than falling — no rule says mediation must produce a resolution in a single meeting.

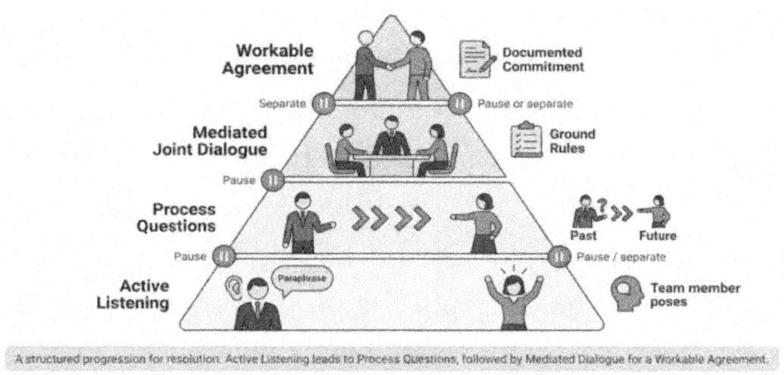

9.9 Building Team Resilience Through Adversity

Resilience is not the absence of difficulty — it is the demonstrated capacity of a team or individual to engage with difficulty productively and emerge with capabilities and relationships meaningfully intact, or even strengthened. This distinction matters because it reframes the managerial goal. The goal is not to create a team that never experiences conflict, disruption, or adversity — that team does not exist outside of very unusual circumstances, and even if it did, it would likely be systematically underdeveloping the conflict navigation and change adaptation capabilities that will inevitably be required later. The goal is to create a team with the relational foundation, psychological safety, communication practices, and leadership support to navigate hard things effectively when they inevitably arrive.

The manager's contribution to team resilience operates through three primary mechanisms. The first is modeling: demonstrating, through your own behavior during difficult periods, that hard things can be engaged with rather than avoided, that setbacks are informative rather than catastrophic, and that the appropriate response to conflict or disruption is structured engagement rather than either emotional reaction or protective withdrawal. The second is structure: providing frameworks, processes, and practices — the team norms, the conflict-resolution channels, the structured retrospectives, the one-on-ones — that give the team reliable mechanisms for processing difficulties rather than having to improvise under pressure. The third is acknowledgment: explicitly naming what the team has been through, validating that it was genuinely difficult, and articulating what the team demonstrated in navigating it.

Post-conflict and post-crisis retrospectives are one of the most consistently underutilized resilience tools in a manager's toolkit. After a period of significant interpersonal tension, organizational disruption, or project failure, a structured reflection exercise — facilitated by the manager, with all parties present, and oriented toward learning rather than blame — performs two organizational functions simultaneously. It converts raw, often painful experiences into organizational learning: what did we actually experience, how did we handle it, what would we do differently if we faced a similar situation, and what can we acknowledge about ourselves in how we got through it? And it creates a shared narrative of survival — 'We went through something genuinely hard together, and we are still here, and we are better positioned to handle the next hard thing because of what we learned in this one' — that is among the most powerful team cohesion statements available.

The relationship between psychological safety and resilience is bidirectional and compounding. Teams with high psychological safety — where speaking up is safe, where admitting a mistake does not carry organizational penalty, where raising a concern is valued rather than suppressed — engage more readily and effectively with conflict precisely because the relational risk of engagement is lower. That more effective engagement produces better outcomes reinforces the team's confidence in its capacity to navigate difficulty, further deepening psychological safety. The manager's task is to maintain conditions for this cycle to operate: modeling vulnerability, rewarding candor with visible responsiveness rather than defensive justification, addressing concerns without retribution, and consistently refusing to let the team's culture be defined by its worst moments.

Diagram 8.8: Team Resilience Building Cycle

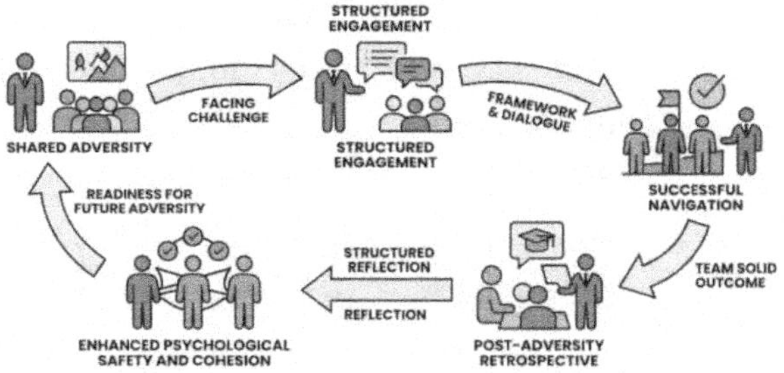

A consistent leadership presence throughout a cycle of shared challenges, structured learning, and reflection strengthens team bonds and prepares groups for future adversity.

9.10 Manager's Checklist — Conflict and Difficult Conversations

Use this checklist as a recurring operational reference. The behaviors listed here are not aspirational ideals. They are the specific, concrete practices that distinguish managers who intentionally shape team culture from those who manage it reactively after the damage is visible.

- Conduct a team charter or working agreement exercise with your team — document norms for disagreement, decision-making, and accountability
- Review team norms quarterly — identify gaps between stated norms and actual team behavior and address them explicitly
- Audit your team's structure for conflict breeding conditions — role ambiguity, competing incentives, resource scarcity — and address root causes before they generate interpersonal friction
- Monitor one-on-one conversations for early conflict signals — frustration with a colleague, feeling dismissed, concerns about process — and act on them while still in the prevention zone
- When conflict requires active intervention, speak with each party individually before facilitating any joint conversation
- Use SBI structure when describing behavior in a conflict intervention context — stay in observable facts and impact, not character assessment
- Do not delay performance conversations — every week of delay makes the eventual intervention harder and more formal
- Prepare for every difficult conversation with both informational clarity and emotional groundedness before you enter the room

- In AI adoption resistance situations, diagnose the specific type of resistance before choosing your response — there is no single-intervention solution
- Use active listening and process questions as your first de-escalation tools — hear before you attempt to resolve
- Conduct a structured retrospective after every significant conflict, disruption, or crisis — convert the experience into explicit team learning
- When managing up, frame your concern as a problem to solve together rather than a complaint to be processed
- Know your organization's escalation channels for concerns that require HR, legal, compliance, or ethics involvement

9.11   The Takeaway

Conflict is not the opposite of a healthy team culture. It is, when handled well, evidence of one. Teams that never have visible disagreements are not psychologically safe environments — they are psychologically suppressed ones, where the norms around raising concerns are so constraining that people with genuine concerns stay silent. The goal is not an environment free of conflict. The goal is an environment in which conflict surfaces early, gets addressed directly and fairly, and is used to clarify what the team values, strengthen the relationships between its members, and improve the processes and structures that shape how it operates. That environment does not emerge by accident. It is deliberately and consistently built by managers willing to have the conversations most people would rather not have.

The difficult conversation you are currently postponing — the underperformance issue you have been hoping will self-correct, the interpersonal friction you have been documenting rather than addressing, the disagreement with your own manager that you have been expressing only in private — is almost always less damaging to address directly than to continue avoiding. The direct address, prepared and grounded, yields a defined, manageable outcome you can build on. The continued avoidance produces an outcome that compounds silently until it becomes a crisis, orders of magnitude harder to address. Every manager makes this choice continuously, across hundreds of small moments. The pattern of those choices defines what kind of leader you are.

Build your conflict competence the same way you build any consequential skill: with deliberate practice, progressive challenge, and honest reflection after each attempt. The first truly difficult conversation you handle well will feel like an achievement significant enough to remember. The tenth will feel like a learned reflex. The hundredth will feel like the professional capability it is — one that your team depends on you to have, and one that sets the foundation for everything else you are trying to build. A manager who can navigate conflict with skill and composure, who can hold team relationships together through turbulence while also advancing the work, is not a rare kind of leader. They are a developed one.

10 Time Management and Personal Productivity in the AI Era

10.1 A Scenario Worth Sitting With

By eight-thirty on a Tuesday morning, Amara had already responded to eleven Slack messages, reviewed a draft report her team had sent at midnight, been tagged in three separate email threads about an upcoming client presentation, received a meeting invite that overlapped with two meetings already on her calendar, and started a task she would not get back to until Friday — if then. She had been a manager for four months. She had not had a sustained, focused block of thinking time in three full weeks. The sensation that she was always in motion but rarely progressing — that she was responsive but not actually managing — had become her persistent professional state. On a good day, she felt useful. On most days, she felt like a very expensive scheduling interface.

What Amara was experiencing is the central, underacknowledged paradox of early management: the role demands far more of your sustained attention, pulls you in far more directions, exposes you to far more incoming information, and grants you far less protected thinking time than being an individual contributor ever did — at the exact moment when your decisions carry more organizational consequence than they ever have before. The job requires strategic thinking, people judgment, and deliberate, context-rich choice-making. The environment in which it is performed rewards reactivity, context-switching, rapid responsiveness, and the visible performance of busyness. If you do not actively and continuously work against this mismatch, the environment wins every time. The result is a manager who is technically present across all their responsibilities and genuinely effective in almost none of them.

This chapter is about building the operational infrastructure to lead effectively rather than react indefinitely. It covers the manager's time audit and what it actually reveals, prioritization frameworks that hold up under real managerial conditions rather than collapsing when the first unexpected urgent request arrives, the specific and hard-won tactics for protecting strategic thinking time in a role that is structurally hostile to it, and the suite of AI tools that can genuinely reduce the administrative overhead that consumes manager bandwidth — provided they are deployed with intention rather than added to the existing pile of things that demand attention. Personal productivity is not a self-improvement sideline topic. It is a foundational management prerequisite. A manager who cannot manage their own time, energy, and attention cannot reliably manage a team. Leadership requires presence, and presence requires space that has to be consciously created and vigorously defended.

Balanced focus requires intentional management of conflicting priorities.

10.2 Why This Matters

The manager who cannot manage their own time cannot manage a team's time, and this is not a judgment about personal discipline — it is a description of an observable organizational reality. Your calendar communicates your priorities more precisely and more credibly than anything you say about them. The manager whose calendar is entirely reactive — filled wall-to-wall with meetings they did not initiate, constantly interrupted by notifications they have not filtered or organized, perpetually behind on commitments that required focused thinking — is not accidentally modeling ineffective work practices. They are actively demonstrating to their team, through revealed behavior, what the organization apparently values and what professional habits it rewards. If you want your team to protect time for focused work, to be deliberate about their priorities, to be present in the conversations they are in rather than perpetually half-distracted by the next incoming notification — you have to model exactly those behaviors yourself.

The personal productivity stakes for managers are substantially higher than for individual contributors because of the nature of the decisions they make and the organizational surface area those decisions affect. As an individual contributor, most of your professional decisions affect the quality and timeline of your own work. As a manager, your decisions affect your direct reports' day-to-day experience, their development trajectories, and their sense of being valued and supported. They affect the team's resource allocation, its strategic direction, and its relationships with peer teams and senior stakeholders. A strategic judgment made in the three minutes between consecutive meetings, distracted by the previous conversation and anticipating the next one, operating without access to the full context the decision requires, is a qualitatively different decision than the same judgment made in a protected block of focused thinking with relevant information assembled and the cognitive space to work through the trade-offs. The gap between those two versions of the same decision is not trivial. Over time, it shapes whether your team operates with strategic coherence or in a perpetual state of responsive improvisation.

The AI productivity opportunity for managers is real, meaningful, and accessible today. Tools for scheduling automation, administrative task generation, meeting summarization, inbox management, and workflow coordination can collectively reduce the administrative overhead that consumes manager time by meaningful percentages — some managers who have systematically implemented them report reclaiming eight to twelve hours per week that had previously been absorbed by low-value but time-consuming administrative work. But the opportunity operates on a specific and easy-to-violate logic: AI tools reduce the time cost of certain administrative and informational tasks, which creates time budget that was not previously available. Whether that budget is actually used for higher-value management work or simply flows back into more reactive behavior — more meetings accepted, more Slack threads responded to immediately, more time spent on the next tier of administrative work that the AI did not address — depends entirely on the manager's own intentionality and structural discipline. The tools create the space. The manager decides what to put in it.

10.3   The Manager's Time Audit — Where Your Hours Really Go

Before you can make intelligent, evidence-based decisions about how to reallocate your time, you need an accurate picture of where it is currently going. This sounds obvious, but it is consistently one of the most revelatory exercises managers undertake, precisely because most managers' beliefs about their time allocation are significantly inaccurate. Research on professional time tracking consistently finds that people underestimate the time they spend on email, meetings, and administrative coordination by thirty to fifty percent, while overestimating the time they spend on high-value, strategy adjacent activities by similar margins. The manager's time audit closes this self-assessment gap — and it is uncomfortable precisely because the gap it closes is substantial.

A rigorous manager's time audit tracks every working hour for at least two full calendar weeks and categorizes each block of time into one of four buckets. Operational time is spent keeping current commitments on track, managing immediate deliverables, and resolving day-to-day issues that arise in the normal course of team activity. Administrative time is spent on scheduling coordination, email and messaging management, producing or reviewing status reports, processing approvals, and maintaining the procedural machinery of organizational life. Strategic time is spent on planning, stakeholder relationship building, team and organizational development, forward-looking analysis, and the thinking required to make consequential decisions well. Personal growth time is spent on the manager's own learning, skill development, reflection, and recovery. The goal of the audit is not to reach a predetermined ratio but to create an accurate baseline — to see, clearly and specifically, what your time is actually producing versus what you intend it to produce.

The patterns that emerge from honest time audits are remarkably consistent across managers, industries, and organizational levels. Administrative tasks — email, scheduling, status meetings, approval queues — consistently consume thirty to forty-five percent of the average manager's week. Operational work accounts for another 30 to 40 percent. Strategic and development activities, which most managers name as their highest-priority responsibilities and which organizations consistently report as the most underprovided management behaviors, receive less than 15% of available time in most cases. Personal growth receives even less. The gap between managers' intended time allocation and their actual allocation is among the most reliable findings in the management research literature. The time audit makes this gap personal and precise, which is what makes it actionable.

Once you have a two-week baseline, the next analytical step is to categorize each time bucket by two dimensions: value created by this activity and substitutability — could this be handled by someone else, automated by a tool, batched with similar activities to reduce total overhead, or eliminated without meaningful consequence? Activities that are high-value and low-substitutability — one-on-ones, difficult performance conversations, strategic planning sessions, stakeholder relationship management, key hiring decisions — belong in the protected, unhurried, peak energy category. Activities that are low-value and high-substitutability — drafting routine status reports, scheduling coordination, compiling standard metrics for recurring reviews, processing straightforward approval requests — are your primary AI automation and delegation targets. This analysis, conducted honestly and with a willingness to follow where the data leads, reveals a substantial reallocation opportunity in virtually every case.

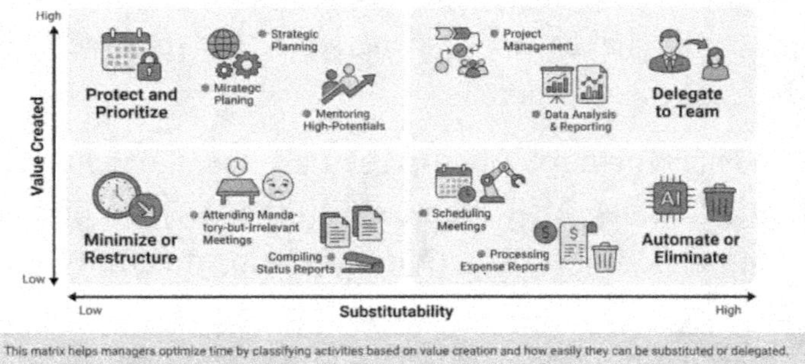

10.4 Prioritization Frameworks — Eisenhower, Time Blocking, and Energy Management

The Eisenhower Matrix is the most widely recognized prioritization framework in the management canon, and its durability is deserved: it provides a simple, intuitive mental model for making priority decisions quickly and consistently. The core logic divides all tasks into four quadrants along two axes — urgency (does this require action soon?) and importance (does completing this create significant value, or does failing to complete it create high cost?). The prescription is straightforward: Urgent and Important tasks get done immediately. Important but Not Urgent tasks get scheduled deliberately with protected time. Urgent but Not Important tasks get delegated or minimized. Neither Urgent nor Important tasks get eliminated or deprioritized. For managers, the most organizationally dangerous of these quadrants is the third — Urgent but Not Important — because the culture of responsiveness in most professional environments continuously manufactures urgency, pulling activities from the fourth quadrant into the third through nothing more than framing and timing.

The limitations of the Eisenhower Matrix in real managerial application are worth naming explicitly, because they reveal where the framework needs to be extended. The matrix treats tasks as binary — either urgent or not, either important or not — when in practice urgency and importance exist on spectrums that interact with each other and with contextual variables in complex ways. More significantly, the matrix does not account for cognitive cost, required emotional availability, or the specific conditions that different types of management work require to be done well. Delivering a performance conversation and drafting a project milestone update may both fall into the 'schedule it' quadrant under a strict Eisenhower analysis. Still, they require fundamentally different cognitive and emotional states to execute effectively. Stacking them back-to-back, or scheduling the performance conversation for the end of a day when you are cognitively depleted, is a prioritization failure that the matrix alone cannot prevent.

Time blocking addresses the scheduling dimension of prioritization that the Eisenhower Matrix leaves implicit. Time blocking means pre-allocating specific blocks of your calendar to specific categories of work before others can fill them with meeting requests, before incoming demands can crowd out your intentions, and before the gravitational pull of reactive work can consume the day. The manager who has blocked Monday mornings for strategic analysis, Tuesday and Thursday afternoons for one-on-one conversations, and Friday mornings for the week's writing and review work is operating with a fundamentally different degree of intentionality than the manager with an open calendar who is perpetually available to whatever request arrives first. Time blocking does not create rigidity — it creates a defended structure within which flexibility operates. The difference is that you are making active choices about what flexibility serves, rather than being flexible by default about everything.

The third prioritization dimension — energy management — adds the biological reality that the Eisenhower Matrix and time blocking both treat as fixed, but is actually highly variable across the day and week. Cognitive performance is not constant. Research on circadian rhythms and cognitive performance consistently finds that most people have a peak cognitive window — typically two to four hours per day, most commonly in the morning or early afternoon but varying significantly by individual chronotype — during which their capacity for focused, creative, and analytically demanding thinking is at its highest. Outside of that window, performance on cognitively demanding tasks degrades meaningfully, even when subjective effort and time invested remain constant. The implication for managers is straightforward: schedule your most cognitively expensive work — strategic planning, difficult personnel decisions, high-stakes stakeholder communications, performance review writing, new frameworks and analyses — in your peak energy window. Schedule cognitively lighter work — email triage, recurring meetings, standard approvals, scheduling coordination — in your lower-energy periods. This scheduling alignment, practiced consistently, produces better work in less time.

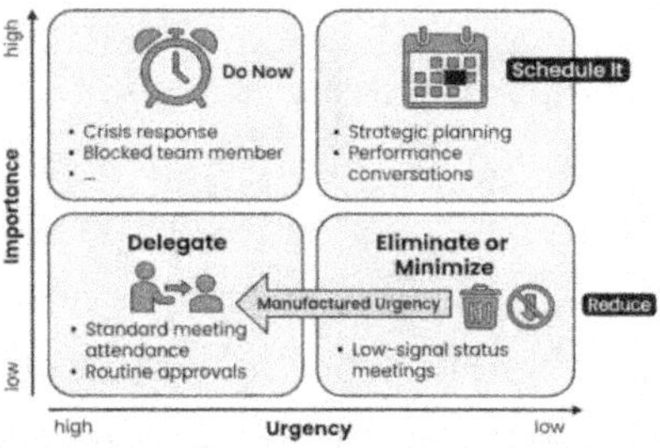

Diagram 9.3: Eisenhower Matrix for Managers

10.5   Taming the Meeting Monster — Fewer, Shorter, Better

Meetings are the single most significant vehicle through which managerial time is consumed, and managerial effectiveness is either supported or undermined. The average professional in a knowledge work organization spends more than thirty percent of their workweek in meetings. For managers at the middle and senior levels, the figure is typically 50 to 60 percent, with significant upward variation in meeting-heavy organizational cultures. What makes this sobering is not the absolute number. Still, when examined critically, what it represents: a significant proportion of that time — organizational research consistently estimates 30 to 50 percent of meeting time — produces no discernible outcome that requires a synchronous meeting to generate. Meetings that exist to share information that could have been a well-structured document. Meetings that exist because they are recurring and have never been formally canceled. Meetings that include twelve people when three of the twelve actually need to be there, and the other nine are present because the meeting culture treats attendance as a proxy for engagement.

The starting diagnostic question for any meeting is simple and consistently clarifying: what outcome does this meeting need to produce, and is a synchronous meeting the most effective and efficient way to produce it? Outcomes that genuinely benefit from synchronous meetings include real-time collaborative problem-solving where the solution is unknown and requires iterative exchange, decisions that require substantive debate among people with different perspectives and organizational authority, relationship-building interactions that require the social and emotional signals of real-time conversation, and conflict resolution conversations where tone and body language carry meaning that asynchronous text cannot reliably convey. Outcomes that do not require meetings include information sharing that can be documented and read on each recipient's schedule, status updates that can be captured in a shared project management tool, decisions that one person has the authority and information to make independently, and feedback that can be delivered more specifically and thoughtfully in writing.

As a manager, you have greater authority over your team's meeting culture than most new managers do. You can establish a team norm of twenty-five- and fifty-minute meetings as defaults rather than thirty- and sixty-minute, creating buffer time that prevents the cascade of late arrivals throughout the day and provides transition time for cognitive reset between consecutive conversations. You can require that any non-recurring meeting request to you be accompanied by an agenda — a specific list of what will be discussed or decided, by whom, with what background the attendee should have — before you accept it. You can establish at least one meeting-free half-day per week for your team, protecting a block of focused work time that benefits everyone. You can audit the recurring meetings you own, cancel the ones whose outcomes could be served by a shared document or an async update, and redesign the ones that remain to have clearer agendas, tighter attendee lists, and explicit decisions documented at the end.

AI meeting assistants have become genuinely capable tools that address one of the most persistent and time-consuming meeting inefficiencies: the overhead of note-taking, action item capture, and post-meeting communication. Tools like Otter.ai, Fireflies.ai, Microsoft Copilot integrated into Teams, and Google's meeting intelligence features in Workspace can transcribe meetings in real time, generate structured summaries of what was discussed and decided, extract specific action items with the names of the people responsible for each one, and distribute all of this to participants within minutes of the meeting's conclusion. The operational consequence of this capability is substantial: meeting notes that used to take twenty to forty minutes to write now take two minutes to review and send. Action-item accountability that previously relied on whether individual participants remembered what they committed to is now automatically documented and traceable. The manager who has not yet deployed meeting AI into their workflow is leaving a significant time-recapture opportunity on the table.

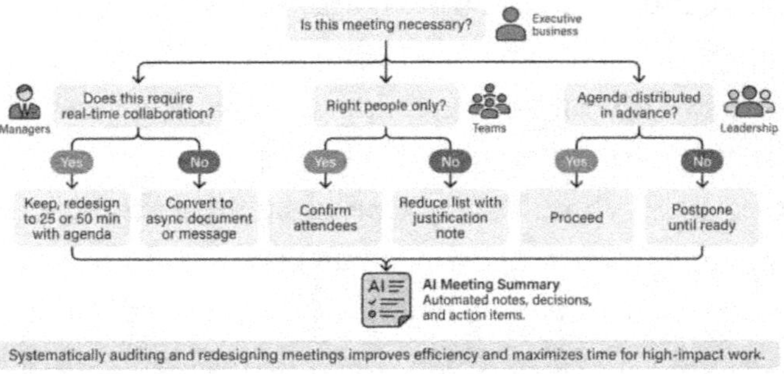

Diagram 9.4: Meeting Audit and Redesign Decision Tree

10.6   AI Scheduling Assistants and Calendar Optimization

Calendar management — the ongoing activity of scheduling meetings, responding to availability requests, managing conflicts among competing obligations, rescheduling when priorities shift, and maintaining a calendar that reflects strategic priorities rather than whoever made the requests first — consumes an enormous proportion of managers' bandwidth relative to the organizational value it creates. Studies of professional time allocation consistently find that executives spend three to four hours per week on pure scheduling coordination. For managers with larger stakeholder networks or organizations where scheduling culture is deeply fragmented, the figure is frequently higher. This is time spent on work whose primary characteristic is that it is necessary without being interesting, consequential without requiring the manager's specific judgment, and perfectly suited to automation.

AI scheduling tools have matured to the point where they can genuinely handle most scheduling coordination on the manager's behalf. Scheduling platforms like Calendly eliminate most of the back-and-forth in external meeting booking by presenting a real-time availability interface that the other party can use to book without exchanging emails. More sophisticated tools like Reclaim.ai and Motion go substantially further. Reclaim.ai analyzes your calendar in real time, automatically reschedules lower-priority blocks when higher-priority conflicts arise, protects designated focus blocks from meeting encroachment, and sequences your meetings to minimize context-switching — for example, clustering similar meeting types together and avoiding isolated thirty-minute blocks that are too short for focused work but long enough to disrupt the flow of a full morning. Motion creates a fully dynamic daily plan that integrates your calendar, task list, and project deadlines, automatically re-sequencing tasks and appointments when new items or delays are introduced.

The principle that makes AI calendar management genuinely useful rather than merely convenient is this: your calendar should reflect your priorities, not just your availability. Without intentional design, most managers end up with calendars that reflect other people's priorities — whoever was first to send a meeting invite, whoever had the most urgent-feeling request, whoever in the organizational hierarchy sent an invitation that felt socially impossible to decline. AI calendar tools, configured correctly, can enforce your own priority structure on that dynamic. They can ensure that the strategic planning block you have designated as protected does not get absorbed by a meeting that could have been async. They can arrange the day's obligations in an order that aligns your highest-cognitive-demand work with your peak energy window. They can create and maintain the scheduling infrastructure that most managers know they need, but cannot do so manually under the volume of competing requests they receive.

The investment required to configure AI scheduling tools effectively is the most common reason managers underestimate their potential. Deploying a scheduling AI with its default settings and expecting it to optimize productivity is like deploying a new team member without onboarding and expecting them to immediately perform at a senior level. The tool needs to know what your actual priority hierarchy is: which commitments are truly non-negotiable, which meeting types belong in which time blocks, how much buffer time you need between consecutive meetings to transition effectively, what your peak energy window is and therefore when strategic thinking time should be scheduled, and what activities you are committed to protecting from encroachment regardless of what else is on your calendar. Investing 30 to 60 minutes in an explicit configuration session upfront — treating it with the seriousness of onboarding a new operational system — typically yields several hours per week within the first month.

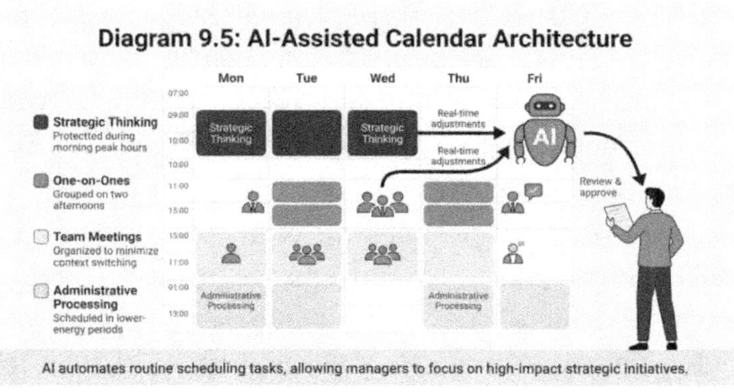

10.7 Automating Administrative Tasks — Reports, Status Updates, Approvals

A substantial proportion of managerial administrative work is structurally repetitive — it follows a consistent format, draws on predictable and accessible data sources, and could be produced at equivalent or higher quality by an AI tool that understood the template and had access to the relevant inputs. Status reports. Weekly team updates. Sprint summaries. Project milestone communications. Standard approval responses for common request types. Performance review summary drafts. These are all candidates for AI-assisted generation or workflow automation, and the aggregate time and cost of producing them manually — over the year and across a career — is substantial. The opportunity cost of not addressing them is not just the time to produce the output; it is the strategic and people development work that could occupy the manager's attention if the administrative overhead were reduced.

The most accessible AI automation pathway for most managers begins with large language model tools —

ChatGPT, Claude, Microsoft Copilot, Google Gemini, or their enterprise integrated equivalents — as drafting assistants. A well-constructed prompt, combined with the relevant data inputs — the sprint completion metrics, the project milestone status, the weekly team accomplishments — can produce a polished draft of a status report, stakeholder update, or performance review summary in under a minute. The draft requires your review, your judgment about accuracy, your calibration of tone for the specific audience, and your addition of context or nuance that the data inputs did not capture. But the mode shift from composing to editing is a meaningful productivity gain. Editing a draft that is 80% of the way to completion typically requires one-third to one-fifth of the time that composing the same document from scratch would require, with no reduction in quality and often an improvement, because the editing eye catches what the composing mind misses.

Workflow automation tools — Zapier, Make (formerly Integromat), Microsoft Power Automate, and similar platforms — address a different category of administrative overhead: the manual handoffs between the software systems that managers and their teams operate in daily. A workflow that automatically posts a team Slack notification when a project milestone is marked complete in your project management system eliminates the need to manually cross-post the update. A workflow that compiles the week's completed tasks from your project management tool and drafts a status report for your review every Friday morning eliminates the need to gather and organize that information manually. A workflow that creates an approval task in your personal queue when a team member submits a PTO request through the HR system, and archives it with the outcome once processed, eliminates both the manual tracking and the risk that requests fall through the cracks during a busy week. These automations are not individually transformative, but collectively, across a week, they eliminate a meaningful volume of the small, repetitive, cognitive overhead-carrying tasks that accumulate into significant time costs.

For managers in organizations with enterprise AI platform licenses — Microsoft 365 Copilot integrated across Word, Excel, Outlook, and Teams; Google Workspace with Duet AI across Docs, Sheets, and Gmail; Salesforce Einstein across CRM workflows; ServiceNow's AI capabilities across IT and operations workflows — the automation potential extends into more sophisticated territory. Intelligent inbox triage that categorizes, prioritizes, and drafts responses for common message types. Document generation from meeting transcripts and rough notes. Contract and policy document summaries that surface the key terms a manager needs to know without requiring them to read every clause. Dashboard and report auto population from operational data streams. The specific capabilities depend on your organization's tool stack, AI licensing, and data governance policies. Still, the directional opportunity is consistent across platforms: any recurring administrative task that follows a predictable format and draws on data that an AI tool can access is a candidate for meaningful acceleration or full automation.

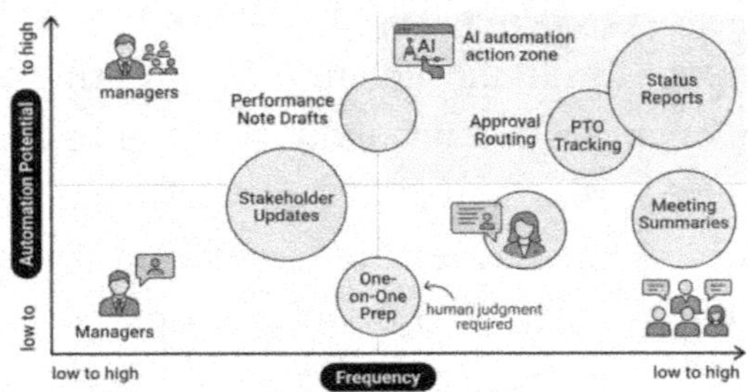

10.8　Email and Communication Management with AI Filtering and Drafting

Email remains, despite the proliferation of team messaging platforms, one of the largest and most persistently unmanaged time sinks in a manager's professional week. Research on professional communication patterns consistently finds that knowledge workers spend twenty to twenty-eight percent of their total working time on email management. For managers who receive more email than their direct reports by virtue of being on more distribution lists, receiving more escalations, and being stakeholders in more organizational processes, the proportion is often higher. The email problem, however, is not fundamentally a volume problem — though volume is real. It is an attention fragmentation problem. Each email notification, if engaged with immediately upon arrival, costs not just the time to read and respond but the six to twenty-three minutes of recovery time required to return to full concentration on whatever task was interrupted. A manager who continuously receives and responds to email throughout the day is not managing their inbox. Their inbox is managing them.

AI-powered inbox management operates at two levels that have meaningfully different implications for manager productivity. The first level is filtering and triage: AI tools learn from the manager's behavior — which emails get opened and responded to quickly, which get archived without reading, which trigger multi-paragraph responses versus brief acknowledgments — and use those patterns to surface the highest-priority messages first, batch lower-priority messages for periodic review, and in some cases automatically handle entirely predictable message types without requiring manager attention at all. Gmail's Priority Inbox has offered a basic version of this capability for years. More sophisticated current tools — Superhuman's AI prioritization features, SaneBox's inbox organization, enterprise AI capabilities now being integrated into Microsoft Outlook and Google Workspace — provide substantially more refined prioritization, can generate suggested draft responses for common message types, and can surface action items buried in long email threads without requiring the manager to read every message in full.

The second level is AI-assisted drafting, which addresses the composition cost rather than the triage cost of email management. For managers who receive significant volumes of email requiring substantive but relatively formulaic responses — project status inquiries from stakeholders, approval requests from team members, meeting requests requiring scheduling coordination, standard policy clarification questions — AI drafting tools can generate a full response draft that the manager reviews, refines, and sends, rather than composing from scratch. The time savings from draft-and-edit versus compose from scratch are consistent and substantial, particularly for managers who communicate regularly across multiple stakeholder groups with different context requirements and communication expectations. The AI draft gives you a complete, coherent starting point; your review and editing process make it sound like you, reflect your actual judgment about the situation, and catch any factual errors or contextual misalignments that the AI introduced.

Beyond email specifically, the broader communication management challenge for managers involves simultaneous streams of Slack or Teams messages, project management tool notifications, calendar alerts, and the various automated notification streams that flow from the operational systems a modern team depends on. Attempting to maintain real-time awareness of all of these streams simultaneously is neurologically impossible and professionally counterproductive — the cognitive overhead of continuous monitoring degrades the quality of every other task performed throughout the day. An evidence-based communication management system for managers involves three structural disciplines. First, configure notification settings to surface only genuinely high-priority signals in real time — typically direct messages from your own manager and project critical alerts — while batching everything else for deliberate periodic review. Second, establish two to three defined communication check-in blocks per day of twenty to thirty minutes each, during which you process accumulated email and messages with full attention, rather than treating communication as a continuous background task. Third, configure AI-powered communication tools — the digest and summary features now available in Slack, Teams, and email platforms — to give you a five-minute summary of the key signals from your communication channels, rather than requiring you to scan hundreds of individual notifications manually.

Diagram 9.7: Email and Communication Management System

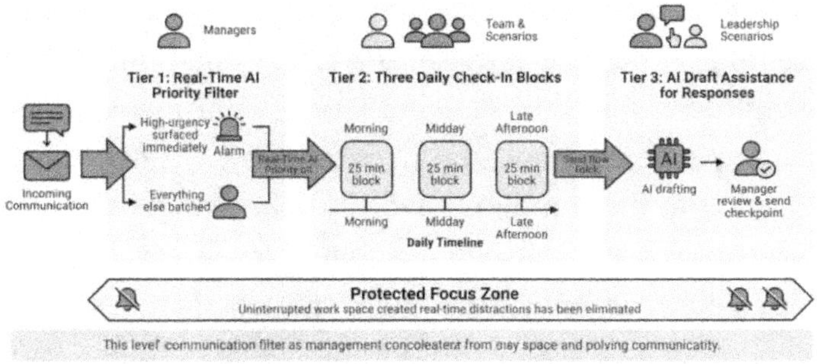

10.9 Protecting Strategic Thinking Time in a Reactive Role

The most critical and hardest-to-maintain element of the AI-era manager's productivity architecture is protected strategic thinking time. This is the time during which you do the work that only you can do — thinking through a difficult personnel decision with full knowledge of the individuals, their histories, their strengths, and the organizational dynamics at play. Working through the implications of a strategic resource allocation choice for your team's mission and their long-term development. Preparing for a high-stakes stakeholder conversation with the depth of preparation that makes the difference between a reactive and a strategic engagement. Reflect honestly on your own leadership patterns, what is working and what is not, and what you want to do differently in the week ahead. No AI tool does this work for you. The AI tools reduce the administrative overhead that previously crowded this work out — but you have to actually use the time that is created, and use it deliberately, for the work that requires the fully present, fully resourced version of you.

The structural mechanics of protecting strategic thinking time are straightforward, and the behavioral discipline required to sustain them is significant. Block time on your calendar before others can fill it, and treat those blocks with the same inviolability that you give to meetings with senior organizational stakeholders — because in terms of long-term organizational impact, they are more important. Label them in a way that communicates both to you and to others that they are genuine commitments: not 'Free' or 'Tentative,' which signals availability, but 'Strategic Planning' or 'Focus Block' or 'Manager Working Session,' which signals intentional use. Communicate the existence and purpose of these blocks to your direct reports so they understand when you are in a focused mode and can defer non-urgent needs to a later time rather than interrupting. And when something urgent-feeling arrives during a strategic block, apply a deliberate test before abandoning the block: Is this more important than my strategic effectiveness? The honest answer to this question is, in most cases, no.

The content of strategic thinking time deserves as much intentional design as the protection of the time itself. Walking into a protected block without a clear agenda for the thinking you intend to do is almost as unproductive as not having the block at all — the default, without a clear question to engage, is to drift toward the mental equivalent of email, checking and rechecking information rather than producing new thinking. A useful practice is to identify and write down, at the beginning of each strategic block, the specific question you most need to think through clearly to lead your team well in the coming period. The question might be a personnel decision, a priority trade-off, an assessment of why a team process is consistently underperforming, or a communication strategy for a difficult stakeholder conversation. Writing the question focuses attention, establishes a definition of productive use for the block, and provides a clear output target. By the end of this block, I should have a well-reasoned answer to this specific question.

Weekly and quarterly reflection practices compound the value of individual strategic blocks by creating a meta-level awareness of your own management effectiveness over time. A brief weekly reflection — fifteen minutes, consistently protected on Friday afternoon or the equivalent, never cancelled for an administrative task — that addresses three questions (what worked well this week and why; what did not go well and what would I do differently; what is most important to prioritize in the coming week) builds a personal feedback loop on your management practice that no external assessment can replicate. A quarterly strategic review that examines your goals, your team's development progress, your time-allocation data from a new audit, and your relationships with key stakeholders provides the longer-horizon perspective that the daily urgency of the management role continually competes against. These practices require no technology, no external resource, and no organizational approval. They require only the discipline to protect time, which, as this entire chapter has argued, is the fundamental management skill on which everything else depends.

A proactive schedule that enables consistent focus and goal achievement.

10.10 Manager's Checklist — Personal Productivity and AI Tools

Review and update this checklist quarterly, along with a new time audit. Personal productivity systems degrade naturally under the pressure of organizational demands unless actively maintained. What works in month three may need redesign by month nine.

- Conduct a two-week time audit every quarter — compare actual allocation against intended priorities and identify the largest gaps
- Categorize your time by value created and substitutability — identify your primary AI automation and delegation targets

• Block strategic thinking time on your calendar before others can fill it — treat it as non-negotiable

• Block all one-on-ones, team rituals, and development conversations as protected recurring commitments that survive schedule pressure

• Establish at least one meeting-free half-day per week for your team — protect it against encroachment consistently

• Require agendas for all non-recurring meeting requests before accepting them

• Audit your recurring meetings quarterly — cancel any where an async document could serve the outcome

• Configure AI scheduling tools with your actual priority hierarchy, protected blocks, and peak energy window — invest the setup time

• Implement AI-assisted drafting for at least three recurring document or email types — establish a draft then edit workflow

• Set up workflow automation for at least three recurring administrative handoffs this quarter — start with the highest-frequency tasks

• Turn off real-time notifications for all but the highest-priority channels — schedule three deliberate daily communication check-ins

• Protect a fifteen-minute weekly reflection block every Friday — what worked, what did not, what matters most in the week ahead

• Complete a quarterly strategic review of your goals, team development, and time allocation — adjust all three based on honest assessment

10.11 The Takeaway

Personal productivity is not a peripheral, aspirational concern for managers who have handled everything else. It is the operational system on which all other management capabilities run, and its absence degrades every other dimension of leadership effectiveness. The manager who is chronically overwhelmed, perpetually reactive, and cognitively depleted by administrative overhead cannot give their team members the quality of attention that genuine development requires. Cannot bring the caliber of thinking that consequential strategic decisions deserve. Cannot enter difficult conversations with the groundedness and preparation that produce good outcomes. Cannot model the professional behaviors — focus, intentionality, sustainable pace, healthy boundaries — that their team members need to observe to develop them in themselves.

The AI productivity opportunity that is now available to managers is real and material. Scheduling tools that protect your priorities against organizational entropy. Drafting assistants who convert composition work into editing work. Workflow automation that eliminates the repetitive administrative handoffs that consume time without creating value. Meeting intelligence tools that make the post-meeting follow-up faster and more reliable. These capabilities exist, they work, and the managers who have deployed them thoughtfully report meaningful time recapture that is now going toward the management work that actually requires it. But none of them produce outcomes without the foundational discipline of knowing where your time is going, being honest about the gap between that reality and your intended priority structure, and protecting the high-value space that automation creates.

Begin with the time audit. It is the hardest step because it is the most honest one. Most managers who have done it report being surprised — sometimes genuinely startled — by the accuracy of the gap it reveals between their beliefs about their time allocation and the measurable reality. That gap, made visible and specific, is the most precise possible prescription for where to start. A few protected blocks, a few automated workflows, a few recurring meetings converted to shared documents, a configured AI scheduling tool — these structural changes, implemented deliberately and maintained consistently, compound into dramatically different professional patterns over months and years. You are not the first manager to feel overwhelmed by the scale of what this role demands. You will not be the last. But the managers who build deliberate productivity systems early in their management careers create the operational foundations on which sustained effectiveness is built, and they lead differently for it — not just in how much they produce, but in the quality of presence they bring to everything they do.

11 AI Fundamentals Every Manager Must Know

11.1   When the Pitch Sounds Too Good

A software vendor walked into your conference room last Tuesday. They brought slides, a live demo, and a confident claim: their AI platform would reduce your team's manual reporting time by 70% within 90 days. The room leaned in. One of your directors started nodding. The procurement contact began scribbling notes. You sat at the head of the table feeling a familiar discomfort — not skepticism exactly, but something quieter and more dangerous. You were not sure what questions to ask.

That discomfort is where this chapter begins. You do not need a computer science degree to lead a team that uses AI tools. You do not need to understand gradient descent, transformer architectures, or the mathematics of neural networks. What you do need is a working mental model of how AI systems behave — their genuine strengths, their structural limitations, and the failure modes that do not appear in demo environments but surface reliably in production. Without that mental model, you cannot evaluate a vendor claim, catch a flawed AI recommendation, protect your team from compliance exposure, or make the call that matters most: when to trust the machine and when to override it.

The discomfort you felt in that conference room was not ignorance — it was the appropriate signal of an informed mind encountering incomplete information. The vendor's demo was optimized for a controlled scenario using pre-selected data and a curated workflow. Your environment is messier, your data is more fragmented, your team's skills are varied, and your regulatory context is industry-specific. Sixty-eight percent of AI projects that never reach production — a figure consistent across multiple enterprise research surveys — fail not because the technology is bad, but because the gap between demo performance and operational reality was never measured, let alone closed.

This chapter is a manager-level primer, not a technical tutorial. Every concept is explained through the lens of the decisions you actually make. By the end, you will know enough to hold a vendor accountable, ask the right questions of your data team, navigate your organization's AI governance conversations, and build a principled framework for evaluating any AI tool that lands on your desk. That is the baseline every leader in the AI era needs — and it is more achievable than most vendors or educators suggest. You are not starting from zero, and you are not alone in this journey.

Diagram 10.1 The Manager AI Knowledge Gap

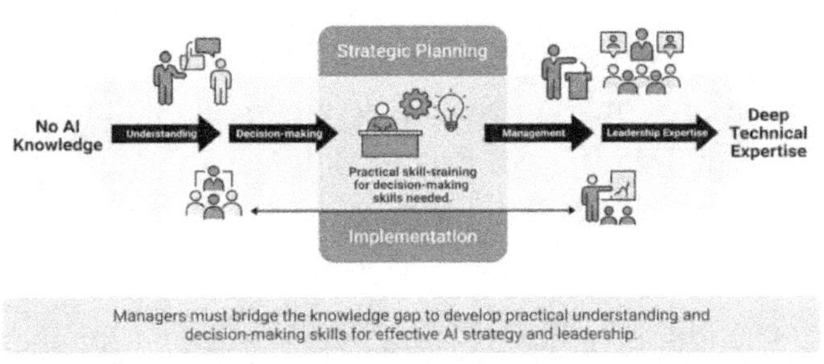

11.2   Why AI Literacy Is Now a Core Management Competency

There was a time when a manager could reasonably say, 'I leave the technology decisions to IT.' That time is over. AI is no longer confined to a specialist function or a back-office system. It sits inside the productivity tools your team uses daily — inside email clients that suggest responses, inside project management platforms that flag schedule risk, inside HR systems that screen resumes and identify flight risk, inside the dashboards where you review team performance metrics. If you cannot read those outputs critically, you are not leading — you are ratifying whatever the algorithm decided.

The organizational risk is direct and measurable. AI systems make errors in patterned, often invisible ways. A language model that writes fluently but fabricates facts. A classification model that performs well in aggregate but fails systematically on a particular demographic subgroup. A forecasting model trained on pre-pandemic data that misreads current demand signals. These are not exotic failure modes cooked up by researchers in controlled settings — they appear in real deployments at real companies, with real consequences for real people. The managers who caught these failures did so because they understood enough about how the underlying systems work to ask the right question at the right moment. The managers who missed them signed off on outputs that looked authoritative on a dashboard.

The opportunity is equally concrete. Managers who develop AI literacy become translators — people who can move fluently between technical teams that build systems and business stakeholders who need results. That translation function is scarce, increasingly valuable, and not easily automated. Your value equation as a manager rises sharply when you can take an AI output, explain what drove it, identify its confidence boundaries, and communicate its implications to a non-technical audience. That is not a technical skill requiring specialized training. It is a leadership skill built on a foundation of technical understanding — and the foundation is the part you build in this chapter.

There is also a talent-and-culture dimension that managers underestimate. Your team members are watching how you engage with AI. If you treat every AI output as authoritative and self-validating, you are modeling uncritical adoption. This behavior creates operational risk and suppresses the skeptical thinking that catches errors before they cause harm. If you dismiss AI tools out of unfamiliarity or discomfort, you signal that growth and experimentation are not team values, and you lose ground to teams that are building capability. The middle path — engaged, curious, appropriately skeptical, and structurally rigorous — is the one that builds a team culture of responsibility by design. That culture starts with your own fluency, demonstrated consistently in how you talk about AI, ask about it, and decide about it.

Finally, consider the career dimension for yourself. The managers who navigate the AI era most effectively are not those who delegate all AI decisions upward to a chief AI officer or sideways to an IT team. They are the ones who maintain sufficient working knowledge to participate meaningfully in AI strategy conversations, evaluate vendor proposals rigorously, advocate for governance structures that protect their teams, and model the balanced, evidence-based engagement with technology that the next generation of workers needs to see. AI literacy is not a credential you earn once and keep for life. It is a practice you maintain as the landscape evolves.

11.3 AI Demystified: A Working Mental Model

11.3.1 Machine Learning: Pattern Recognition at Scale

The term artificial intelligence covers a wide range of technologies, and the breadth of what gets called AI in vendor pitches and industry coverage creates genuine confusion. For managers, the most practically important category to understand is machine learning — a class of systems that learn patterns from historical data rather than following explicit rules coded by a programmer. Understanding this distinction matters because it changes how you should think about errors, reliability, and the appropriate scope of trust.

A traditional software system does exactly what it is programmed to do. If the rules are wrong, the output is wrong in a predictable, traceable way — you can read the code and find the error. A machine learning system does something fundamentally different: it identifies patterns in a

273

large dataset and applies those patterns to new inputs it has never seen. If the training data is unrepresentative, biased, or outdated, the model will produce outputs that reflect those problems — and the errors can be entirely invisible until they cause harm in the field. The system is not broken in the traditional sense; it is doing exactly what it was trained to do, on data that did not adequately represent the full reality you needed it to model. That distinction between traditional software errors and machine learning errors is one of the most important conceptual tools a manager can develop.

This has direct management implications that play out repeatedly in practice. When you evaluate a machine learning-based tool, the question is not only 'does it work in the demo?' The more important questions are: What data was it trained on? Does that data represent my specific context, my team, my customers, and my operating environment? What happens when the real world diverges from the training distribution — as it inevitably will? These are not highly technical questions requiring a data science background. They are data literacy questions — and every manager who has evaluated a vendor contract or reviewed a business case can learn to ask them systematically.

Three subcategories of machine learning appear most frequently in management contexts, and recognizing them helps you match the right questions to the right tool. Supervised learning trains a model on labeled examples — for instance, past employee performance reviews paired with eventual retention outcomes — and uses those patterns to predict future cases. Unsupervised learning identifies natural groupings in data without predefined labels, used in customer segmentation, document clustering, and anomaly detection. Reinforcement learning trains systems through reward signals, most visible in recommendation engines, dynamic pricing algorithms, and game playing AI. You do not need deep expertise in any of these subcategories. You need to recognize them when a vendor or data team describes their system, ask what labeled data or reward signal was used to train it, and probe whether that training setup reflects the outcomes you actually care about.

11.3.2 Generative AI: A Different Kind of System

Generative AI has reshaped the management landscape faster than almost any prior technology category. Large language models — the technology behind tools like ChatGPT, Microsoft Copilot, Claude, and Google Gemini — operate on a fundamentally different principle than traditional predictive machine learning. They are trained on massive corpora of text to predict what word, phrase, or sentence is most likely to follow a given input. The outputs are often remarkably coherent, useful, and even creative. They are also structurally prone to a failure mode that every manager must understand deeply before deploying them in a professional context: fluent incorrectness.

A large language model does not retrieve facts from a verified database. It generates text that is statistically consistent with its training data. When it produces a confident-sounding statement, that confidence is a linguistic property of the output — a reflection of how sentences are structured in the training corpus — not a signal of factual accuracy. This is why language models hallucinate, producing plausible-sounding but incorrect facts, nonexistent citations, fabricated statistics, and legally problematic claims. The model is not lying in any meaningful sense. It is completing a pattern with the words most statistically likely to follow. For managers using these tools to draft communications, summarize reports, or generate analysis, the practical implication is always the same: verify factual claims before distributing any output. Generative AI accelerates drafting enormously; it does not replace verification, and it must not be treated as doing so.

Generative AI also operates in modalities beyond text, and each carries its own risk profile. Image generation tools create visual content from text prompts, raising questions about copyright ownership of the training data and of the generated outputs — questions that remain legally unsettled in most jurisdictions. Audio and video synthesis tools can produce realistic speech and moving images of real people, creating potential for misinformation, brand risk, and misuse that managers must account for in acceptable use policies. Code generation tools translate natural language descriptions into executable software, accelerating development but introducing security vulnerabilities when the generated code is not reviewed by engineers who understand its implications. The manager's role is not to evaluate these risks at a technical level but to ensure that someone within the team or the organization's governance structure has explicit ownership of each risk category before the tool is deployed.

A useful mental model for generative AI in management contexts is to treat these tools as highly capable first-draft colleagues who have read extensively but verify nothing independently and have no stake in the accuracy of the output. That framing keeps the workflow right: AI contributes the draft, the human contributes the judgment, the verification, and the accountability. When organizations invert that model — treating AI output as the final product and human review as optional — they create conditions for the kinds of visible, embarrassing, and sometimes legally consequential errors that generate headlines and destroy trust.

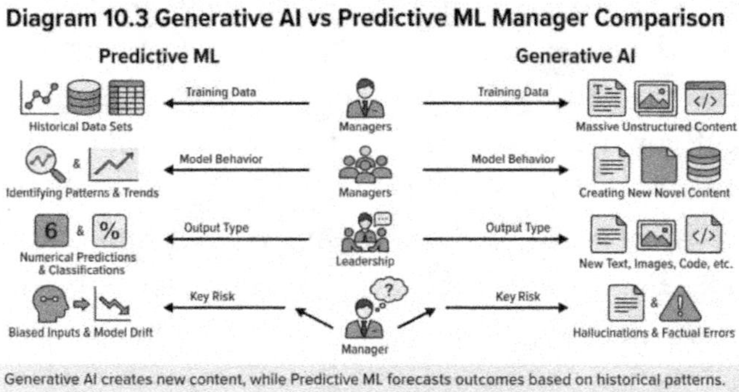

11.3.3 Natural Language Processing: The Bridge Between Humans and Systems

Natural language processing (NLP) is the branch of AI concerned with enabling machines to understand, interpret, and generate human language. It underpins many of the AI tools managers encounter daily: sentiment analysis in your employee survey platform, summarization in your meeting transcription tool, entity extraction in your contract review software, intent detection in your customer service chatbot, and topic modeling in your market research analysis. Understanding NLP at a conceptual level helps you calibrate expectations for each application and identify failure modes before they reach stakeholders.

The most important NLP concept for managers is context sensitivity. Language carries meaning that is deeply and often subtly context-dependent. The phrase 'we need to move on this' can convey urgency, closure, or transition, depending on the surrounding context, the speaker's tone, the organizational culture, and the relationship between the parties. NLP systems vary significantly in their ability to handle ambiguity, specialized domain vocabulary, irony, cultural reference, and multilingual content. When you deploy an NLP-based tool — particularly one that analyzes employee sentiment or processes customer feedback — the first validation question should always be: was this model tested on text that looks like our text? A consumer sentiment model trained on product reviews will not reliably interpret internal Slack messages, technical support tickets, or the specific idioms of your industry.

NLP is also the technology most directly implicated in the content policies and acceptable use decisions your organization needs to make. Language models can generate inappropriate content, reproduce private information from their training data, or be prompted in ways that bypass safety filters through what are called jailbreaking techniques. These are not hypothetical concerns — they have appeared in production deployments at well-resourced organizations. They are governance questions that fall within a manager's purview: setting behavioral expectations for AI-assisted work on the team, establishing review protocols for AI-generated content before it reaches external audiences, and fostering a culture where team members feel empowered to flag AI outputs that seem inappropriate or inconsistent with organizational values.

11.4   The AI Tool Landscape: Categories That Matter to Managers

Rather than cataloging individual vendors — a task that would be obsolete before this page is printed, given the pace of the market — the more durable approach for a manager is to understand the functional categories of AI tools and what each category reliably does and does not do well. This framework travels across vendor changes and product updates, giving you a stable lens for evaluating new tools as they emerge.

Automation tools handle defined, rule-based tasks without human intervention. Robotic process automation, which predates modern AI by many years, belongs in this category. So do AI-enhanced versions that add document classification, data extraction, intelligent routing, or simple decision branching to the automation layer. These tools work best on high-volume, structured, predictable tasks where the exception rate is low and the cost of an unhandled exception is manageable. They fail at tasks that require judgment, handling exceptions beyond a narrow threshold, or contextual interpretation of ambiguous inputs. When a vendor claims their automation tool can handle 'any document' or 'any workflow,' the right response is: what percentage of real-world documents or workflow instances from a deployment similar to yours fell outside the tool's automated capability in your most rigorous pilot testing?

Augmentation tools amplify human capability without removing the human from the loop. Writing assistants, meeting summarizers, code completers, research assistants, and design tools fall into this category. These have the highest manager adoption rates and the most immediate and visible productivity payoff. They are also the tools where uncritical acceptance creates the most risk, because the quality of the AI output is often high enough to suppress the verification instinct. An AI writing assistant that produces a polished, professional first draft can create the illusion of completeness. The manager's role in governing augmentation tools is to establish team norms around verification, attribution, and quality review — norms that treat AI-generated content as a capable first draft, not a final product, and that maintain the team's own expertise rather than allowing it to atrophy through dependence on AI.

Analytics and insight tools process data to surface patterns, predictions, and recommendations. Sales forecasting, attrition risk modeling, project risk detection, customer segmentation, financial anomaly detection, and supply chain optimization all belong in this category. These are the tools most likely to influence high-stakes business decisions — and therefore the tools where data quality, model transparency, and calibration rigor matter most. Before relying on an analytics tool's output for a significant decision, a manager should be able to answer three questions without hesitation: What data feeds this model? How recently was the model trained or refreshed? Has it been validated on data from our specific business context, not just a generic industry benchmark?

Conversational and workflow AI tools — including chatbots, virtual assistants, and AI orchestration platforms that route work across systems — handle real-time interactions and coordinate tasks between applications. These are high-visibility tools that customers and employees interact with directly. Their performance is visible and immediate in a way that backend analytics tools are not — failures are experienced as frustration, confusion, or loss of trust, not merely as statistical error rates on a monitoring dashboard. Managing conversational AI tools requires particularly close attention to feedback loops, escalation pathways to human agents, the quality of the human handoff when the AI cannot resolve a case, and the ongoing maintenance cadence for keeping conversation scripts and knowledge bases current.

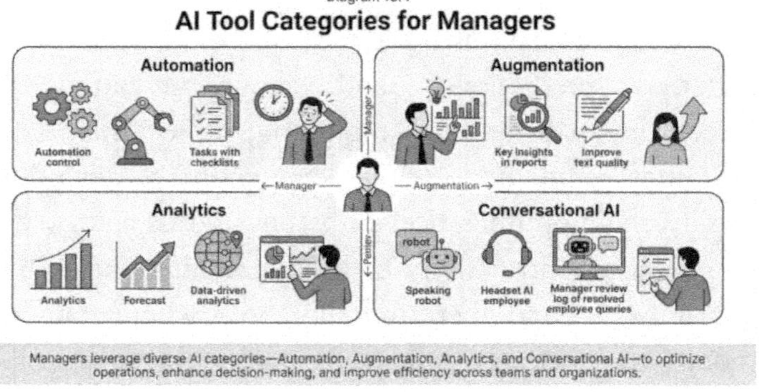

Managers leverage diverse AI categories—Automation, Augmentation, Analytics, and Conversational AI—to optimize operations, enhance decision-making, and improve efficiency across teams and organizations.

11.5   Data Literacy for Managers: Inputs, Outputs, and Limitations

The phrase 'data-driven decision making' has become so pervasive in organizational language that it has nearly lost its meaning. For a manager engaging with AI tools, data literacy means something specific and operational: understanding what data went in, what the output actually represents, and what the system cannot know. These three questions form the foundation for responsible AI use at the management level and apply consistently across tool categories.

11.5.1 Understanding Inputs

Every AI system is fundamentally a product of the data it was trained on. The boundaries of that data define the boundaries of the system's reliable knowledge — not just what the model knows, but what it can know at all, given what it has seen. A demand forecasting model trained on three years of pre-pandemic transaction data has no inherent way to account for the supply chain disruptions, behavioral shifts, or macroeconomic volatility that emerged afterward. It will still produce forecasts — complete with confidence intervals and trend lines — but those outputs are pattern-matching against a world that no longer exists in the same form. The model is not defective; it is extrapolating accurately from a sample that no longer represents the population you need to predict.

For managers, the practical implication of input awareness is that AI outputs must always be evaluated in light of what the training data could and could not see. When your project risk model flags a workstream as low-risk, it is pattern-matching against historical projects with similar structural characteristics. If your current project has features that do not appear in the training history — a new technology stack, a geographically distributed team for the first time, a compressed timeline driven by a regulatory deadline, a vendor relationship without historical precedent — the model's confidence rating is measuring similarity to past patterns, not actual risk level. Your judgment about those novel factors is essential, not optional. The model cannot see what it was not taught to look for.

Input quality encompasses data freshness, completeness, and labeling consistency, each of which can lead to distinct failure patterns. A model trained on employee engagement survey data that has not been refreshed in eighteen months is measuring a workforce that may have turned over significantly, been reorganized, or experienced major cultural events since the training cutoff. A model trained on data labeled by multiple human raters with inconsistent criteria will exhibit inconsistent behavior on similar inputs because the training signal itself was inconsistent. These are questions you should direct routinely at whoever manages your organization's AI systems — and they require no technical background to ask effectively. What is the recency of the training data? Were labeling criteria defined and enforced consistently? What is the plan for data refresh as the operating environment changes?

11.5.2 Interpreting Outputs

AI outputs come in several forms, each requiring a different kind of interpretive discipline. Probability scores — expressed as percentages, confidence intervals, or risk ratings — are the most commonly misread class of AI output. A model that assigns a 68% probability that a project will miss its deadline is not predicting the future with 68% accuracy. It is saying that, in the historical training data, projects with these characteristics had late outcomes 68% of the time. Whether those historical projects are truly comparable to your current one — in team composition, organizational context, technology environment, and stakeholder dynamics — is a judgment call that belongs to you, not to the algorithm. The number is an input to your decision, not a conclusion.

Classification outputs — where a model places an input into one of several predefined categories — carry a different, often underappreciated risk: they present discrete certainty when the underlying reality is continuous. A resume screening model that labels a candidate as 'likely qualified' or 'unlikely qualified' is compressing a nuanced, multi-dimensional evaluation into a binary label. The threshold at which the model switches from one category to the other is a design choice made by the tool's developers, set to optimize a particular metric on a particular dataset. That threshold may not be calibrated to your organization's specific values, role requirements, or team context. Understanding where the threshold sits — and what happens to the candidates or cases that fall close to the boundary — is essential for any decision process where the outcome materially affects a person.

Text and content outputs from generative AI must always be treated as first drafts requiring substantive human review, not finished artifacts ready for distribution. This is not a counsel of excessive caution — generative AI produces genuinely useful output that can dramatically reduce the time required to produce a first draft. It is a recognition of the structural reality that the fluency and polish of the output are not correlated with its accuracy, and that distributing unreviewed AI-generated content creates organizational, legal, and reputational risk that grows in proportion to the sensitivity and visibility of the audience. The standard that applies to any other professional output — that it reflects the judgment and accountability of the person whose name is on it — applies equally to AI-assisted output.

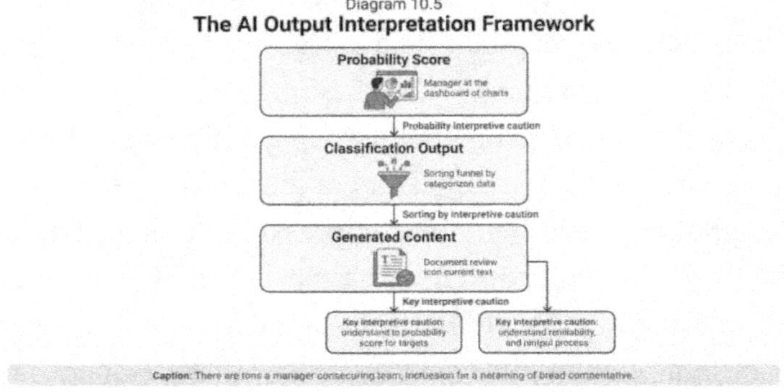

Caption: There are tons a manager consecuring team, incruesion for a netaming of bread compentative.

11.6   AI Ethics, Bias, and Responsible Use

AI bias is not an edge case or an aspirational concern reserved for conference panels and academic papers. It is a documented, recurring, and measurable phenomenon that affects deployed systems in healthcare, hiring, credit, criminal justice, and enterprise productivity tools. For managers, this is not primarily a philosophical topic — it is a risk management topic with direct legal, operational, and human consequences. Deploying or relying on a biased AI system exposes your organization to discrimination claims, regulatory violations, reputational damage, and — most fundamentally — to producing worse outcomes for the people your work is supposed to serve.

Bias enters AI systems through multiple pathways, and understanding those pathways helps you ask the right questions at the right stage of an AI adoption process. Historical bias occurs when training data reflects past human decisions that were themselves biased — a hiring model trained on ten years of promotion decisions will replicate whatever patterns of advancement existed in the organization, including demographic disparities that resulted from structural inequality rather than differences in merit. Representation bias occurs when the training dataset underrepresents certain groups, leading the model to perform worse on those populations because it has learned less about them. Measurement bias occurs when the data used to define and label outcomes is itself flawed — a performance rating system with known rater consistency problems creates a contaminated foundation for any AI model trained to predict or assess performance.

Managers do not need to audit the mathematics of a model's bias or conduct their own statistical validation. They do need to ask four governance questions of any AI system that affects people's opportunities, evaluations, access to services, or career trajectories. First: has this system been tested for differential performance across demographic groups, and what were the results? Second: Who within the vendor organization is accountable for ongoing bias monitoring and remediation after deployment? Third: What is the escalation path within our organization if team members or affected individuals observe outcomes that appear discriminatory? Fourth: Does our legal and compliance team have visibility into exactly how this tool is being used and what decisions it informs? These questions are governance questions, and asking them systematically is a leadership behavior, not a technical one.

Responsible use of AI also requires intellectual honesty about the relationship between AI systems and the decisions they inform. When a manager uses an AI system to inform a performance rating, a hiring decision, a resource allocation, or a customer treatment, the manager remains the decision-maker of record. An AI tool cannot be named as the responsible party for choices that affect people's lives and livelihoods — not legally, not ethically, and not in a way that produces accountability within an organization. This principle has practical implications for how you design AI-assisted workflows: build them to surface AI as a contributor to the decision, not as a replacement for the decision-maker. The human review step is not a formality — it is the point where organizational accountability is exercised. Remove it, and you have diffused accountability into an algorithm that cannot be held responsible for anything.

The growing regulatory environment makes responsible AI use both a compliance imperative and a moral one. Regulators in multiple jurisdictions have made it clear that 'the AI made the decision' is not an acceptable explanation for a discriminatory or harmful outcome. Organizations are expected to understand how their AI systems work, test them for harmful bias, monitor them in deployment, and maintain human accountability for their consequences. The manager who builds these practices into their team's AI workflows before a compliance audit is not just doing the right thing — they are building the operational infrastructure that protects the organization from the class of AI-related liabilities already materializing in enforcement actions globally.

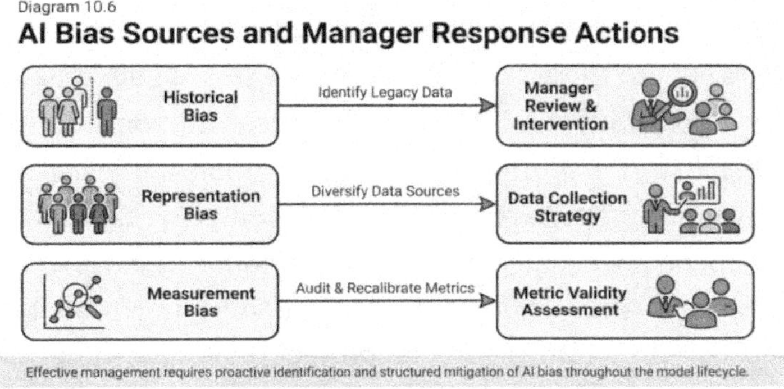

11.7 Privacy, Security, and Compliance Considerations

Every AI tool that your team uses processes data. In most enterprise contexts, that data includes information about customers, employees, business partners, financial performance, and strategic plans — data categories that are simultaneously valuable to the organization and sensitive in their implications for the people they describe. The intersection of AI capability and data sensitivity creates a compliance surface that managers cannot afford to treat as someone else's problem. The legal and IT functions play critical roles. Still, decisions that expose an organization to data risk often happen at the team level, in everyday choices about which tools to use and what data to feed into them.

The central privacy question for any AI tool is a simple one that managers should ask before any new tool enters team practice: when data goes into this system, where does it go, how long is it stored, by whom, under what access controls, and can it be used to train or improve future versions of the model? These questions are not always answered clearly in standard vendor agreements, and the answers are often buried in terms of service documents that are dense with legal language but clear in their implications when read carefully. Many AI tools — particularly consumer-grade applications that employees may adopt independently and informally without IT review — use submitted data to improve their underlying models. That means your team members' project plans, customer names, product specifications, internal financial projections, or strategic planning documents could become training inputs for a model that serves other organizations' users. This is a material data governance risk, and it begins with the adoption decisions made at the team level, not with a central IT policy that arrives months after the practice is established.

From a security perspective, AI tools introduce new attack surfaces that managers need to be aware of, even if they do not manage security directly. Prompt injection — where a malicious actor embeds instructions inside content that an AI system will process — can cause AI tools to reveal confidential information, produce harmful outputs, or take unintended actions within integrated systems. AI-generated synthetic content, including deepfakes of real individuals, creates social engineering risks that are more sophisticated and more scalable than traditional phishing. AI models that have direct access to internal data systems through API integrations can become the vector for unauthorized data exfiltration if not properly scoped, monitored, and permissioned. Managers do not need to become security engineers. They do need to ensure that any AI tool requesting access to internal systems has passed through the organization's established security review process before team adoption, not after the first security incident.

The regulatory environment surrounding AI is expanding and becoming more specific. In the European Union, the AI Act creates binding requirements for high-risk AI applications, including systems used for employment decision-making, creditworthiness assessment, and critical infrastructure management. These requirements include transparency obligations, human oversight mandates, data governance standards, and conformity assessments before deployment. In the United States, sector-specific guidance from the Equal Employment Opportunity Commission, the Consumer Financial Protection Bureau, and state-level privacy statutes create accountability standards that extend to AI-assisted decisions in employment, consumer credit, and healthcare contexts. The fact that a vendor sold and configured the tool does not transfer the compliance obligation from the organization using it. As the manager deploying a tool within your team's workflow, you are a responsible party for its compliant use — and that responsibility requires proactive awareness, not just reactive compliance when an incident occurs.

11.8   Building Your AI Evaluation Framework

Every manager needs a repeatable, portable process for evaluating AI tools — whether that tool arrives as a vendor pitch, an IT recommendation, an employee request, an executive mandate, or a competitive intelligence report about what peer organizations are deploying. The goal is not to become a thorough technical evaluator capable of assessing model architecture and hyperparameter choices. The goal is to ask the right questions in the right sequence so that neither hype nor unfamiliarity drives the adoption decision. The following framework applies across tool categories and scales from a quick, informal assessment for low-stakes tools to a structured, multi-week pilot evaluation for high-stakes deployments.

11.8.1 Step One: Define the Problem Before the Tool

The most common and costly AI adoption failure starts with the solution rather than the problem. A team sees an impressive AI demonstration, finds it exciting, and reverse engineers a use case to justify the purchase. The result is a solution looking for a problem — and the problem it eventually finds is rarely the highest-priority one the team faces. Before evaluating any AI tool, invest thirty minutes in articulating in writing the specific workflow problem you are trying to solve: what is the current state, what is the desired future state, what is the measurement that would tell you whether the change worked, and what would the cost be of not solving this problem? This discipline protects against shiny-object adoption and provides a baseline against which any tool's actual performance can be evaluated.

11.8.2 Step Two: Evaluate the Data Fit

Armed with a clear problem definition, evaluate whether the tool's training data and operational data requirements align with your specific context. Ask the vendor directly and in writing: what data was this model trained on, and what are the demographic, geographic, and domain characteristics of that training set? What data does the tool need at runtime to produce its outputs? What happens when required data is missing, inconsistent, or in a format different from what the tool expects? If the vendor cannot answer these questions with specificity, that is a signal — not necessarily that the tool is inadequate, but that the vendor lacks sufficient transparency into its own system to be a trustworthy long-term partner.

11.8.3 Step Three: Assess Explainability and Transparency

Explainability refers to the degree to which a model's outputs can be traced back to specific inputs and reasoning steps in a way that a human decision-maker can understand and communicate. For high-stakes decisions that affect people — performance evaluations, resource allocation, hiring, customer service resolution — you need to be able to explain why the AI produced a particular output. Ask the vendor: for a specific output, can you show me which input features contributed most to that result, and in which direction? Some tools provide this natively via feature importance scores, counterfactual explanations, or natural language rationales. Others are essentially black boxes that produce outputs without any traceable reasoning. The appropriate level of explainability depends on the stakes and regulatory context of the decisions the tool informs — but for personnel decisions, explainability is not optional.

11.8.4 Step Four: Run a Structured Pilot

No AI tool should move from evaluation to production deployment without a pilot that tests performance in your actual context with your actual data on a representative sample of real-world cases. The pilot should include a success metric agreed upon before the pilot begins, rather than reverse engineered from the results, a control condition or baseline comparison where feasible, deliberate testing of edge cases and failure modes that the vendor's demo was unlikely to include, and a structured debrief that evaluates both quantitative performance metrics and qualitative team experience. The pilot-to-production pathway is the moment when vendor claims meet operational reality, and it is the stage at which most AI tool evaluations should end — rather than proceed — if the evidence does not justify confidence in production deployment.

11.8.5 Step Five: Establish Governance Before Deployment

Governance decisions should be made before a tool goes into production, not in response to the first incident that reveals its absence. Who owns the ongoing performance monitoring for this tool? How will errors and anomalous outputs be reported and escalated? What is the explicit process for human override when team members disagree with an AI recommendation? What are the approval requirements for model updates, data refreshes, and configuration changes? Who has the authority to suspend the tool if performance degrades? These questions define the operational accountability structure for AI within your team's workflow — and answering them in advance is a leadership decision that creates the difference between a managed AI deployment and a situation where everyone assumes someone else is watching.

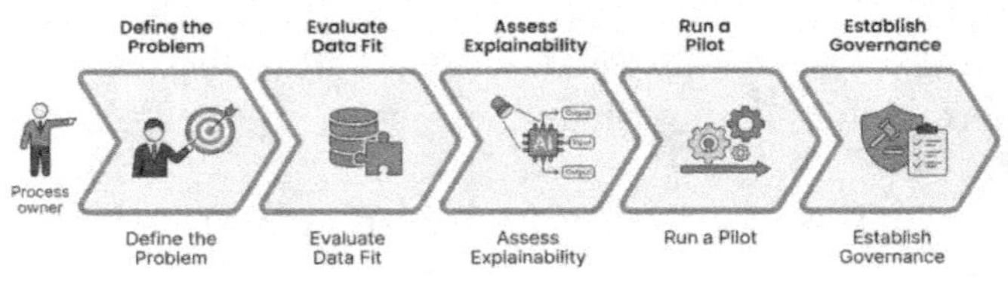

11.9 When AI Helps and When Human Judgment Is Essential

The most operationally valuable judgment a manager can develop in the AI era is the ability to distinguish situations in which AI should inform or accelerate a decision from situations in which human judgment must remain primary and cannot be delegated to an algorithm. This is not a permanent binary that can be settled once and applied forever — the right balance shifts as tools mature, as your team's AI literacy and experience develop, as governance structures solidify, and as the nature of specific decisions changes with business context. But developing a working heuristic saves you from the twin failure modes of AI adoption: under-reliance, which leaves genuine productivity value on the table, and over-reliance, which treats algorithmic outputs as authoritative when they are not.

AI reliably adds value in tasks characterized by high volume, pattern-based logic, large historical datasets, and a tolerance for occasional errors that do not cause irreversible harm. Parsing thousands of customer feedback entries to identify recurring themes. Generating first-draft communications from a template and factual inputs. Scheduling optimization across multiple conflicting constraints. Flagging anomalies in transaction data for human review. Summarizing meeting transcripts and extracting action items. Classifying incoming support tickets by category and urgency. These tasks share a common structure: the AI handles the processing burden at a scale no human can match, and the human applies judgment to the output, adds context the AI cannot see, and takes accountability for the result. The human is in the loop — but their time is concentrated where it creates the most value, rather than on the mechanical processing that precedes the judgment.

Human judgment remains primary — and AI input remains advisory — in tasks involving genuinely novel situations without meaningful historical precedent, high stakes with irreversible consequences for real people, significant ethical dimensions that require moral reasoning rather than pattern completion, interpersonal sensitivity that depends on relationship context the AI cannot hold, or the need to account for organizational and political dynamics that are not captured in any dataset. Performance improvement plans, terminations, compensation decisions, disciplinary actions, strategic pivots in response to new competitive threats, conflict mediation, organizational restructuring decisions, and promotions remain human leadership functions in the full sense. AI tools may provide relevant data and surface patterns the manager might otherwise miss — but the manager who delegates these decisions to an algorithm has abdicated the core responsibility of the role and created the conditions for the kind of documented AI misuse that attracts regulatory scrutiny.

A practical heuristic that holds across decision types: the more that a wrong decision would harm a specific person in a way that cannot be undone, the more clearly human judgment must be primary. AI is excellent at reducing the time and cognitive load required to reach good decisions on reversible, pattern-based, high-volume questions, where occasional errors are manageable. It is structurally unsuited to be the decision-maker for choices that affect individuals' lives, careers, health, and dignity — not because AI is malicious, but because AI cannot be held accountable, cannot be sued, cannot be terminated, and cannot experience the moral weight of the consequences of its outputs. That distinction — between AI-assisted efficiency and AI-replaced accountability — is the moral and operational center of responsible AI leadership in the current era.

11.10 Manager's Checklist: AI Fundamentals in Practice

• Develop a one-page AI literacy primer for your team that explains the categories of AI tools in use, their respective use cases, and their known limitations — review and update it annually.

• Before adopting any new AI tool, complete the five-step evaluation framework: problem definition, data fit assessment, explainability review, structured pilot, and governance design.

• Identify which decisions on your team are currently being informed by AI outputs, and confirm that a specific named human is accountable for each final decision.

• Ask your IT or data team to provide documentation on the training data recency and demographic validation methodology for any AI system that affects personnel decisions.

• Review your team's current use of consumer-grade AI tools and establish a written policy specifying what data categories can and cannot be submitted to those tools.

• Require any vendor pitching an AI tool that affects personnel or customer outcomes to answer in writing: what demographic groups were represented in your validation set, and what were the performance differentials across those groups?

• Establish an AI incident log — a lightweight record of cases where AI outputs were factually incorrect, misleading, or overridden by team judgment — and review it quarterly as a learning tool.

• Schedule a monthly AI transparency check-in with your team to discuss how AI tools are being used, what the team trusts and what they question, and whether current norms remain appropriate.

• Confirm that your organization's legal and compliance teams have reviewed the data handling and privacy terms for every AI tool your team uses that touches customer or employee data.

• Document the human override process for every AI tool that makes a recommendation your team acts on

— ensure every team member knows they have both the right and the professional responsibility to escalate a questionable AI output rather than defer to the algorithm.

11.11 What You Now Carry Forward

The vendor who walked into your conference room last Tuesday was not your adversary — and neither is the technology they were selling. AI tools have genuine, measurable, and growing value in management contexts, and the managers who learn to deploy them well — with clear problem framing, rigorous evaluation, and appropriate governance — will outperform those who do not. But the competitive advantage does not come from adopting the most tools the fastest. It comes from adopting the right tools with the right governance for the right problems, with the right combination of enthusiasm for capability and skepticism about limitations.

The mental model you carry from this chapter is durable and will remain useful as the specific tools in the market evolve rapidly around it. Understand that AI systems learn from data, and data has boundaries that define the system's reliable range. Understand that fluent and confident outputs from generative AI are not correlated with factual accuracy, and that verification is always required. Understand that bias is a structural property of training data that must be tested for and monitored, not assumed away. Understand that compliance responsibility does not transfer to the vendor when the tool is deployed in your team's workflow. Understand that human accountability is not optional in high-stakes decisions affecting individuals, regardless of how sophisticated the AI recommendation appears. And understand that the right question in any AI evaluation is not 'what can this tool do?' but 'what problem am I solving, and is this the most effective and responsible means of solving it?'

AI literacy is not a destination you reach and then stop updating. Every generation of AI tools brings new capability profiles, new failure modes, and new governance challenges that require the same rigorous, principled approach applied to a new set of specifics. The managers who remain effective through successive waves of AI advancement are the ones who treat AI literacy as an ongoing leadership practice — reading critically, piloting carefully, asking hard questions of vendors and themselves, and building teams that share the same habits. You are not the first manager to navigate this terrain, and the path is more navigable than it appears from the vendor's conference room. The foundation you have built here makes every subsequent AI decision on your team more defensible, more effective, and more aligned with the operational clarity that responsible leadership demands.

12 Leveraging AI to Supercharge Team Productivity

12.1   The Bottleneck Nobody Named

Your team is talented. Their technical skills are strong, their motivation is genuine, and each person performs well individually. Yet every sprint ends with a handful of tasks carrying over, every project status meeting surfaces the same recurring delays, and the quarterly retrospective generates the same root-cause themes: too many manual handoffs, too much time spent on low-value coordination work, too many hours lost to searching for information that should be instantly findable. The bottleneck is not skill. It is not an attitude. It is the invisible tax of process friction — the accumulated overhead of repetitive work that AI-assisted workflows can eliminate or dramatically reduce.

Consider a concrete illustration. A mid-sized engineering team spends an estimated twelve percent of its collective working hours each week on work coordination activities: updating project status in multiple systems, reformatting data between tools that do not integrate natively, writing meeting summaries and distributing action items, routing incoming requests to the right team member, and answering recurring questions that are documented somewhere but hard to find quickly. Twelve percent of a ten-person team is more than one full-time equivalent person per week — not creating anything, not solving problems, not building capabilities, and just keeping the coordination machinery running.

This chapter is the practical playbook for recapturing that capacity. It is not about automating people out of existence or replacing human judgment with software routines. It is about building a team environment where AI handles the coordination overhead, the information retrieval burden, and the first-draft drafting work so that your team members spend their cognitive energy on the work only humans can do: creative problem-solving, relationship-building, contextual judgment, strategic thinking, and the kind of domain expertise that generates real organizational value. That reallocation is not a technological project. It is a leadership project — and it starts with how you map, prioritize, and sequence AI adoption across your team's specific workflows.

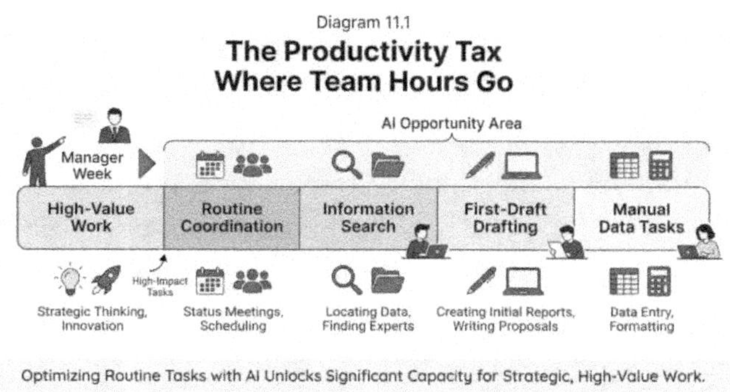

12.2 Why AI Productivity Is a Leadership Mandate, Not an IT Initiative

The conventional mental model of AI adoption places it in the technology department — something IT procures, configures, and deploys, then hands off to business teams to use. That model produces two consistent failure modes. Either the tools that arrive are optimized for technical feasibility rather than workflow relevance, leading to low adoption and shelf-ware. Or the tools that get used are adopted informally and without governance, leading to inconsistent practices, security exposure, and quality variance. Both outcomes leave significant productivity value unrealized while creating operational risk that falls on the manager's desk when something goes wrong.

Effective AI productivity improvement is a leadership mandate because the decisions that determine whether AI tools deliver value are management decisions, not technology decisions. Which workflows are the highest-priority targets for AI assistance? Which team members need capability development to use tools effectively? How do you measure the ROI of an AI tool investment against baseline performance? How do you maintain quality standards when AI is generating first drafts? How do you build a team culture that treats AI as a productivity partner rather than a threat or a novelty? These are people- and process-related questions, and they require a manager's judgment, communication skills, and institutional knowledge to answer correctly in a specific team's context.

The productivity gains available through AI-assisted workflows are substantial and increasingly well-documented. Analysis by consulting firms and technology researchers consistently shows that knowledge workers with access to well-integrated AI tools complete first-draft tasks forty to sixty percent faster, reduce time spent on information retrieval by thirty to fifty percent, and report higher satisfaction with their work mix — spending more time on tasks they find meaningful and less on tasks they find tedious and repetitive. These are not marginal improvements. They represent the kind of workflow-level impact that changes what a team can accomplish in a quarter, how quickly individuals can develop skills, and how effectively a manager can deploy limited headcount against an expanding set of organizational demands.

The risk of inaction is also concrete and accelerating. Teams that build AI-assisted workflows at scale develop genuine organizational capability advantages that compound over time. They complete more work with the same headcount. They build institutional knowledge that is systematically captured and searchable. They reduce the risk of a single point of failure that comes from critical knowledge residing only in individual team members' heads. Teams that delay meaningful AI integration, waiting for a perfect governance framework or an executive mandate that may not arrive, fall progressively further behind their peers in capability and efficiency. For a manager accountable for team output, that gap is a performance problem — and closing it is part of the job.

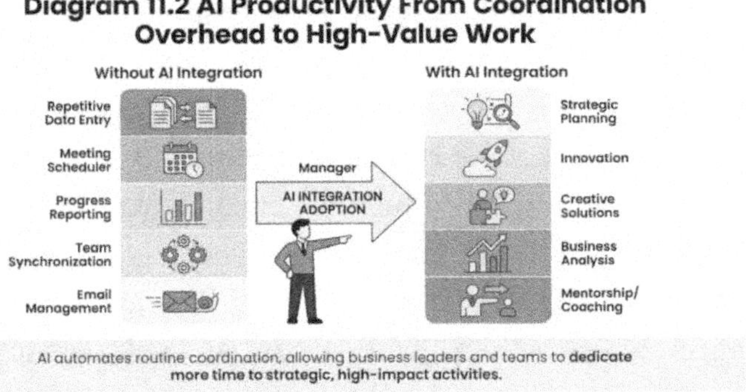

Diagram 11.2 AI Productivity From Coordination Overhead to High-Value Work

12.3   Mapping Your Team's Workflow to Identify AI Opportunities

The most common AI integration mistake is technology first adoption: picking a tool because it is popular or prominently marketed and then looking for places to use it. The more durable approach is workflow first integration:

systematically mapping how your team works, identifying where time is spent on low-leverage activities, and then evaluating which AI capabilities address those specific bottlenecks. This approach produces better ROI, higher adoption rates, and fewer abandoned tool implementations because every tool introduced has a clear, pre-defined job.

12.3.1 The Workflow Audit

Begin with a structured workflow audit — a deliberate effort to map where your team's time actually goes versus where it ideally should go. This does not require sophisticated time-tracking software. A simple instrument works well: ask each team member to track their activities in fifteen-minute blocks for one or two representative weeks, categorizing each block as either value-generating work — output that directly advances the team's objectives and that only a skilled human can produce — or supporting work — output that is necessary but mechanical, repetitive, or primarily concerned with coordination, communication, formatting, and information management. The category boundary will not always be perfectly clean, and that ambiguity is itself informative. When team members are uncertain whether their work is value-generating or supporting, it often signals a workflow design problem.

Aggregate the team's results to identify patterns. Where are the largest concentrations of supporting work? Which activities appear on multiple team members' lists, suggesting systemic rather than individual workflow inefficiency? Which activities involve manual transfer of information between tools or systems that should ideally communicate directly? Which activities require searching for information that the team produces regularly but cannot find quickly? Which recurring tasks involve producing a first draft of standard content — status reports, meeting summaries, responses to common questions — that could be accelerated with AI assistance? Each cluster of answers points toward a specific AI capability category worth evaluating.

The output of the workflow audit is a prioritized opportunity map — a list of workflow segments ranked by three factors: the volume of time currently consumed, the degree to which AI assistance could reduce that time without degrading quality, and the organizational impact of freeing that time for higher-value work. This map becomes your AI integration roadmap. It allows you to sequence AI tool adoption by impact rather than by novelty, and it provides a baseline against which you can measure actual productivity improvements after deployment. This discipline distinguishes serious AI integration from performative adoption of technology.

12.3.2 Identifying High-Value AI Touchpoints

After completing the workflow audit, you will typically find that AI opportunities cluster around four types of activity that appear across team types and organizational contexts. First, information processing tasks: activities that require reading, synthesizing, categorizing, or summarizing large volumes of text, data, or documents. These are the highest-volume AI opportunities and typically the easiest to pilot because success is measurable and failure is visible. Second, first-draft generation tasks: activities that produce written, verbal, or structured output from inputs that are known and defined — status reports, meeting notes, communication templates, documentation, code comments, data summaries. Third, search and retrieval tasks: activities that require finding specific information across multiple systems, documents, or conversation archives. Fourth, coordination and handoff tasks: activities that move work or information from one person, system, or stage to another without adding substantive value to the content being transferred.

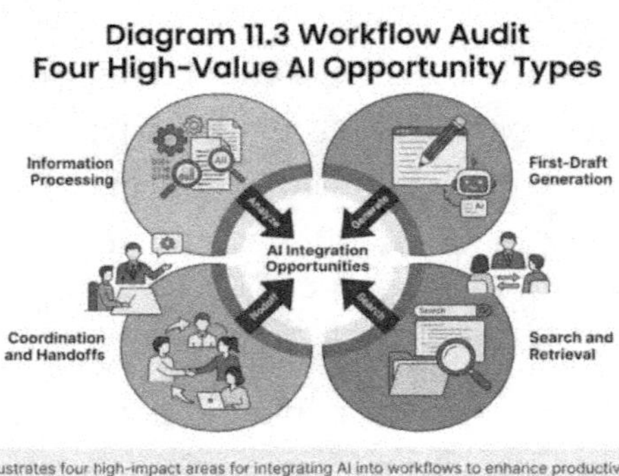

This diagram illustrates four high-impact areas for integrating AI into workflows to enhance productivity and efficiency.

12.4 AI for Project Management: Scheduling, Risk, and Resource Optimization

Project management is one of the most mature areas for AI integration in team workflows, and for good reason: it is a domain rich with structured, historical data — timelines, resource assignments, task dependencies, completion rates, risk events — that AI can analyze to surface patterns invisible to human inspection of individual projects. The AI-assisted project management tools available today operate at several levels of sophistication, and understanding what each level actually does helps you calibrate your expectations and governance requirements accordingly.

At the foundational level, AI-enhanced scheduling tools use historical project data to produce more realistic timeline estimates by correcting for the optimism bias that human estimators consistently exhibit. Studies of software development projects have found that AI-assisted scheduling that incorporates historical velocity data and accounts for complexity factors produces estimates that are 30 to 40 percent more accurate than human-generated estimates without data. This matters operationally because optimistic schedules create cascading downstream problems: missed commitments, reactive resource reallocation, quality shortcuts taken under deadline pressure, and erosion of team morale from consistent underperformance against targets that were never achievable.

At a more sophisticated level, AI risk prediction tools analyze current project characteristics — team composition, dependency complexity, resource utilization patterns, scope-change frequency, and communication patterns in project tools — to identify workstreams at elevated risk of delay or quality issues before those risks materialize as missed milestones. The value of early risk detection is disproportionate: a risk identified three weeks before a milestone can be addressed with moderate effort, while the same risk identified two days before the milestone requires crisis-level intervention and often produces worse outcomes even with maximum effort. These tools are most valuable when integrated into existing project management platforms rather than deployed as standalone applications that require separate data entry, which introduces adoption friction that kills most tool implementations.

Resource optimization tools analyze workload distribution across team members to surface imbalances that create bottlenecks — situations where one team member's queue is the constraint on five other people's work — and to identify utilization patterns that predict burnout risk before team members self-report. For managers managing ten or more team members, manually tracking utilization across multiple workstreams with sufficient granularity to catch these patterns is practically impossible. AI tools that surface the signal amid the noise in task assignment data give managers the visibility needed to make proactive reallocation decisions rather than reactive crisis management decisions.

The governance requirement for all AI project management tools is the same: human review before action. AI-generated risk flags are hypotheses about elevated probability, not predictions of certain outcomes. AI-generated resource rebalancing suggestions require human validation against context that the tool cannot see — personal development goals, team relationship dynamics, specialized knowledge that makes certain assignments uniquely valuable, and organizational priorities that are not captured in task management data. Use AI to surface the signal; apply your judgment to determine the response.

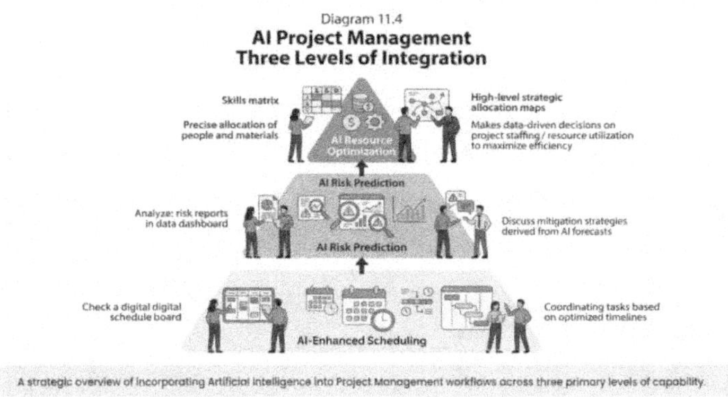

12.5   AI Collaboration Tools: Smart Documents, Co-Authoring, and Meeting Intelligence

Collaboration is the invisible infrastructure of team productivity. Meetings, documents, conversations, and coordination mechanisms are how teams align, make decisions, and transfer knowledge — and they are also where an enormous amount of productive time gets lost to inefficiency. AI collaboration tools address this infrastructure layer directly and represent some of the most immediate productivity gains available to managers who adopt them with a clear purpose and appropriate norms.

12.5.1 Meeting Intelligence and Transcription

Meeting intelligence tools — which combine real-time transcription, speaker identification, action item extraction, and automated summary generation — address one of the most persistent and measurable productivity drains in knowledge work: the overhead of meeting follow-up. The average manager spends an estimated 30 to 45 minutes per meeting on post-meeting tasks: writing summaries, distributing notes, extracting action items, and ensuring accountability for commitments. AI meeting intelligence tools reduce this overhead to minutes without sacrificing quality — and, in many cases, improve it, because AI-generated summaries based on full transcripts are more complete and more accurate than human summaries based on memory and partial notes.

The operational value extends beyond the efficiency of individual meetings. Meeting intelligence tools create a searchable archive of team decisions, commitments, and discussion history that serves as institutional memory — accessible to team members who missed a meeting, to managers onboarding new team members, and to leaders reviewing the evolution of a decision over multiple discussion sessions. This institutional memory function is particularly valuable in hybrid and remote environments where informal knowledge-sharing through hallway conversations is reduced and documented communication becomes the primary shared record.

Implementing meeting intelligence tools requires establishing clear norms with your team before the first recording. Who has access to transcripts? How long are recordings retained? Are there discussion topics where transcription should be suspended — sensitive personnel conversations, protected health information, or legal matters? These norms should be documented, shared with the team, and revisited as the team's experience with the tool grows. Teams that establish clear norms before deployment have significantly higher adoption rates and lower incidence of the trust related complications that can arise when team members feel surveilled rather than supported.

12.5.2 AI-Assisted Document Collaboration

The modern document collaboration stack — epitomized by tools like Microsoft 365 Copilot and Google Workspace AI features — embeds AI assistance directly into the documents, spreadsheets, and presentations where team members spend their working hours. The practical productivity gains from these integrations are substantial when team members are trained to use them effectively rather than discovering them through trial and error. First-draft generation from a brief, structured prompt reduces document creation time by forty to sixty percent for standard business documents. AI-powered summarization condenses long documents and email threads to their essential content in seconds, reducing the reading burden for time-pressured managers. Suggested edits and reformatting recommendations reduce the polish cycle for professional communications.

The governance challenge unique to embedded document AI is the quality review workflow. When AI assistance is seamlessly integrated into the authoring experience, it is easy for team members to accept AI suggestions without substantive review — particularly when under time pressure. The result can be polished-looking documents that are inaccurate or inconsistent in content. Establish an explicit team norm: AI-generated content in any document sent to stakeholders outside the team requires explicit human review and sign-off by the person whose name appears on the document. That norm does not need to slow the work down significantly; it simply ensures that AI assistance accelerates drafting, not accountability avoidance.

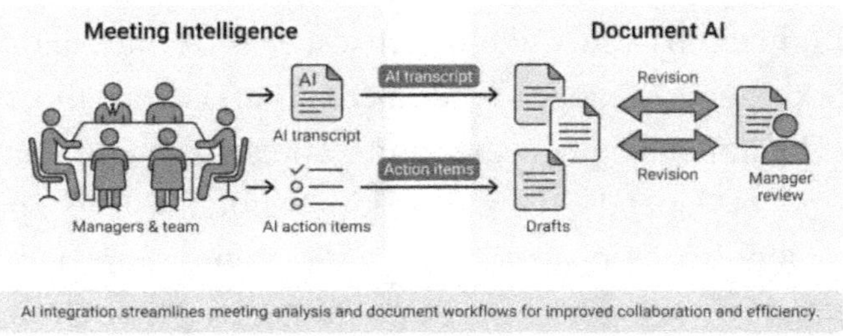

**Diagram 11.5 AI Collaboration Layer
Meeting Intelligence and Document AI**

AI integration streamlines meeting analysis and document workflows for improved collaboration and efficiency.

12.6   Workflow Automation: Eliminating Manual Handoffs

Manual handoffs — moments when work or information is transferred from one person, tool, or system to another through human action rather than automated routing — are among the most significant and fixable sources of delay and quality degradation in knowledge work. Each manual handoff is an opportunity for delay (the work sits in someone's inbox or task queue until they attend to it), error (information is incorrectly copied, partially transferred, or lost in translation between formats), and context loss (the receiving party lacks background that the sending party possessed but did not think to document). Workflow automation — connecting systems and routing work automatically based on defined rules — eliminates these failure modes at scale.

Modern workflow automation tools range from low-code platforms like Zapier and Make that connect hundreds of SaaS applications through pre-built connectors, to more sophisticated enterprise automation platforms like Microsoft Power Automate and ServiceNow that handle complex, multi-step workflows across enterprise systems. For managers, the strategic opportunity is to identify the highest-frequency, highest-impact manual handoff points in the team's workflow and build automation for those points first — rather than attempting to automate everything at once, or pursuing workflow automation as an abstract technical goal rather than a solution to a specific operational problem.

A practical taxonomy of manual handoffs to automate, in rough order of implementation simplicity, includes: notification and alert routing, where information about an event in one system is manually communicated to team members in another channel; data synchronization, where the same information exists in multiple systems and is kept current through manual copying; approval routing, where documents or requests are forwarded through a chain of approvers via email or manual task assignment; and report generation, where data from multiple sources is manually compiled and formatted into a standard document on a recurring schedule. Each of these categories has established automation solutions that most enterprise IT environments can support, and each offers immediate, measurable time savings once implemented.

The management discipline for workflow automation is impact sequencing combined with change management. Automate the handoffs that consume the most time or create the most errors first — the ones that appear on multiple team members' workflow audit results. Involve the team members who currently perform those handoffs in the automation design process, both to capture the full complexity of the current process and to build ownership for the new one. And recognize that automation changes jobs, not just processes: team members whose time is freed from manual handoffs should have clarity about what higher-value work those hours are being redirected toward, rather than discovering that efficiency gain has resulted in downsizing. The productivity benefit of workflow automation is fully realized only when the recaptured hours go toward mission-aligned work, not toward an ambiguous expectation that more will be done with the same inputs.

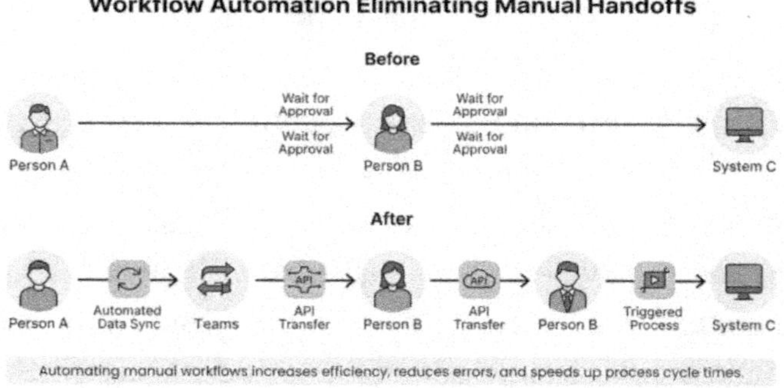

Automating manual workflows increases efficiency, reduces errors, and speeds up process cycle times.

12.7 AI for Knowledge Management: Institutional Memory and Searchable Intelligence

One of the most underappreciated productivity costs in team management is the time spent reinventing knowledge that already exists somewhere within the organization. Team members repeatedly answer the same questions because the answers are not findable. New projects recapitulate research completed by a prior project team because that research was never captured in a searchable format. Onboarding new team members takes months rather than weeks because institutional knowledge lives in individual people's heads rather than in systems that can be efficiently transferred. The organizational cost of poor knowledge management is invisible on a project by project basis but massive in aggregate.

AI-powered knowledge management tools address this problem at two levels. The first is knowledge capture: making it easier to extract and structure knowledge from the places where it is created — meeting conversations, Slack threads, documents, emails, and project deliverables — and store it in formats that are findable and usable. AI transcription and summarization tools automate the most tedious part of knowledge capture, reducing documentation overhead so significantly that team members are more willing to contribute. The second level is knowledge retrieval: making it possible to find relevant information through natural language queries rather than requiring team members to know exactly which folder, system, or person holds the answer.

Enterprise AI search tools — including Microsoft Copilot's enterprise search features, Notion AI, Guru, Glean, and similar platforms — allow team members to ask questions in plain language and receive synthesized answers drawn from across the organization's documented knowledge base. The productivity benefit of this capability is most pronounced in roles where information retrieval is a significant portion of the workday: customer success teams answering client questions, technical support teams troubleshooting complex issues, sales teams preparing for customer conversations, and managers preparing for performance reviews and project status discussions. When team members can get answers in seconds that previously required searching multiple systems for ten to thirty minutes, the compounding effect on team output is significant.

Building a team knowledge management practice requires more than deploying a tool. It requires establishing the behavioral norms that make knowledge capture a consistent team habit rather than an aspiration. The most effective approach is integration into existing workflows rather than the addition of a separate documentation step: meeting intelligence tools that automatically capture and file meeting summaries, project management platforms that automatically archive completed project artifacts, and communication tools that make it easy to save a conversation thread to the knowledge base with a single action. When knowledge capture is frictionless, it happens. When it requires dedicated effort on top of already-full schedules, it does not.

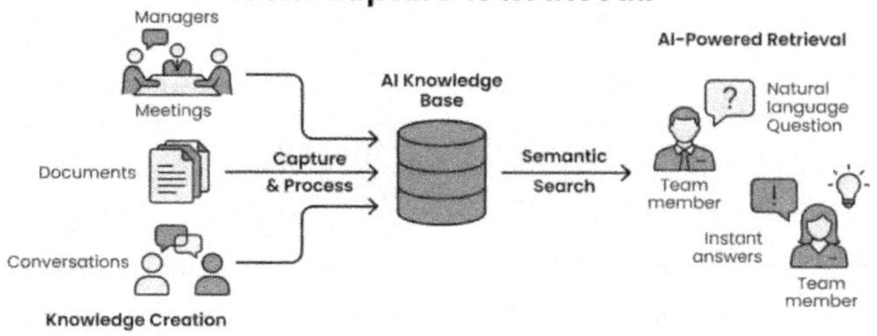

12.8   Measuring ROI on AI Tool Investments

AI tool investments require the same scrutiny as any other operational investment — and that scrutiny requires measurement. The challenge is that AI productivity benefits manifest in ways that are difficult to quantify using standard business metrics. Time savings are the most direct measure but require baseline data to calculate. Quality improvements require before-and-after assessments of output quality. Error-rate reductions require tracking error incidents over time. Employee satisfaction improvements require periodic measurement. None of these is extraordinarily difficult to track, but all require deliberate setup before deployment rather than attempting to reconstruct baselines after the fact.

A practical ROI measurement framework for AI tools operates at two levels. The first is direct productivity measurement: establish a baseline for the specific tasks the AI tool is designed to improve, measure performance after a defined 90-day adoption period, and calculate the time savings per task, multiplied by the task frequency, and multiplied by the fully loaded cost of team member time.

This gives you a direct dollar value for the productivity improvement, which can be compared against the tool's licensing cost to calculate a straightforward payback period. The second level is workflow-level impact measurement: assess whether the team is completing more of its mission-critical work within the same time period, whether project cycle times have improved, and whether team members' self-reported work mix has shifted toward higher-value activities.

Be rigorous about distinguishing adoption metrics from value metrics. The number of users who have logged into a tool or the number of queries submitted to an AI assistant tells you about adoption, not impact. Impact is measured by changes in outputs: faster task completion, higher-quality results, more projects completed per quarter, fewer errors requiring rework, and shorter onboarding time for new team members. Measure adoption as a leading indicator and impact as the lagging indicator that ultimately justifies continued investment or signals the need for course correction.

Communicate AI ROI measurement results transparently to your team. When team members can see concrete productivity data from AI tool adoption — time saved, quality improvements, work completed — it reinforces the value of using the tools effectively. It builds cultural momentum for continued investment in AI literacy. When results are disappointing, transparent measurement allows the team to diagnose the root cause — was it tool selection, adoption approach, workflow design, or governance failure? — and course correct based on evidence rather than assumption. That accountability culture, applied to AI investments as to all others, is what separates mission-aligned AI integration from performative technology adoption.

12.9 Building Your Team's AI Tech Stack: A Decision Framework

The term 'tech stack' originated in software development to describe the combination of technologies that power an application. For a management team, the AI tech stack is the combination of AI tools and capabilities your team uses to accomplish its work. Building this stack deliberately — rather than accumulating tools through a series of independent, uncoordinated adoption decisions — is one of the highest-leverage technology management decisions a modern manager makes.

The first principle of AI tech stack design for managers is coherence over comprehensiveness. A small number of well-integrated, well-adopted tools delivers far more productivity value than a large collection of overlapping tools that team members use inconsistently. Integration matters because productivity gains multiply when tools share data and context — a project management tool that connects to the team's communication tool, which connects to the knowledge management system, which connects to the meeting intelligence tool, creates a connected productivity environment where information flows automatically. Team members do not need to manually maintain the same information in multiple places. Every tool that does not integrate with the rest of the stack requires manual bridging, which recreates the handoff overhead you were trying to eliminate.

The second principle is team capability alignment. The sophistication of the AI tools in your stack should match your team's AI literacy and comfort level, with a modest stretch factor that builds capability over time. A team that has never used AI tools will get more value from a focused deployment of a single well-designed tool with strong training and adoption support than from a comprehensive platform deployment that overwhelms team members and leads to low adoption across all features. Sequence tool introductions to build confidence and capability incrementally — each successful adoption experience builds the team's appetite and readiness for the next level of AI integration.

The third principle is governance before scale. Every tool in the AI stack should have a defined owner responsible for monitoring its performance and user satisfaction, a documented acceptable use policy that specifies what data can be submitted and how outputs should be reviewed, and a clear offboarding process if the tool is discontinued or replaced. These governance elements are easiest to establish when the stack is small and the tools are new. They become progressively harder to retrofit as the stack grows and dependencies deepen. Build the governance infrastructure before you scale the stack, not afterward.

A practical AI tech stack for a typical knowledge worker team today might include: an AI writing assistant embedded in the primary document creation environment, an AI meeting intelligence tool integrated with the team's calendar and communication systems, an AI-enhanced project management platform, a workflow automation layer connecting the team's core SaaS applications, and an enterprise AI search capability connected to the team's knowledge repositories. That is five functional layers — each with a clear job, clear success metrics, and clear governance ownership. That modest, coherent stack, well-adopted and well-governed, will deliver more productivity value than a sprawling collection of twenty tools with inconsistent adoption and no integrated governance.

Diagram 11.8: Manager's AI Tech Stack: Five Integration Layers

Integrating these AI layers empowers managers to streamline operations, enhance collaboration, and make data-driven decisions.

12.10 Manager's Checklist: AI Productivity Integration

- Conduct a structured workflow audit with your team to identify the highest-frequency manual, repetitive, and coordination activities — complete this before evaluating any specific AI tool.

- Prioritize AI integration opportunities by three factors: time currently consumed, feasibility of AI assistance, and organizational impact of recaptured hours.

• Establish a written acceptable use policy for AI collaboration tools before deploying them — specify data categories, review requirements, and escalation paths.

• Before deploying meeting intelligence tools, hold a team conversation about norms: who sees transcripts, how long they are retained, and what discussion topics require transcription suspension.

• Select AI project management features that integrate into your team's existing workflow rather than requiring a separate tool with separate data entry.

• Identify the top three manual handoff points in your team's workflow and assign ownership for evaluating automation solutions for each within a defined ninety-day window.

• Design your knowledge management approach around frictionless capture: embed documentation into existing workflows rather than adding a separate documentation step.

• Establish ROI baselines before deploying any AI tool — measure the current baseline for task completion time, error rate, or output volume that the tool is expected to improve.

• Audit your current AI tool portfolio for integration coherence — identify tools that require manual bridging between them and evaluate whether integration improvements or consolidation are appropriate.

• Build your team's AI tech stack governance structure before scaling: define ownership, acceptable use, and offboarding processes for every tool in the stack.

- Schedule quarterly retrospectives specifically focused on AI tool performance — evaluate adoption metrics, impact metrics, team satisfaction, and whether governance controls are functioning as intended.

12.11 What You Now Carry Forward

The bottleneck your team faces is real and fixable. The coordination overhead, the manual handoff friction, the information retrieval burden, the first-draft drafting work that crowds out creative and strategic thinking — these are not constants of knowledge work. They are workflow design problems, and AI-assisted workflow redesign is now an accessible, practical solution available to managers at every level of technical sophistication. The key is to approach that redesign as a leader: start with workflow analysis, prioritize by impact, sequence adoption to build capability and confidence, measure results rigorously, and govern the tools in a way that keeps human accountability central.

The teams that will define the productivity standard in your industry over the next five years are not the ones that adopted the most AI tools. They are the ones who adopted the right tools, built genuine team capability to use them effectively, established governance structures that maintained quality and compliance, and created a culture of continuous improvement where AI integration is an ongoing practice rather than a one-time implementation project. That culture starts with how you, as the manager, engage with AI productivity tools in your own work — modeling the critical evaluation, governance discipline, and purposeful adoption you expect from your team.

The workflow-level impact of getting this right is compounding. Each AI integration that recaptures time from low-value coordination work and redirects it toward high-value mission-aligned work creates organizational capability that did not exist before. Team members who apply their expertise to expert-level work develop it faster than those who spend it on mechanical processing. Institutional knowledge captured and made searchable creates leverage for every future project, rather than evaporating when individual team members move on. Projects completed more efficiently generate an organizational track record that builds the case for expanded investment and autonomy. That compounding dynamic is the true productivity payoff of AI integration done well — and it is a leadership outcome, not a technology outcome.

13 Data-Driven Decision Making

13.1   The Dashboard That Didn't Tell the Whole Story

The numbers looked unambiguous. Customer satisfaction scores had climbed three points over the prior quarter. First-call resolution rates were up. Average handle time was down. Every metric on the team's operations dashboard was trending in the right direction, and the monthly leadership report reflected exactly that — a clear, data-backed story of a team performing well. Then a senior client called to say their experience over the past ninety days had been the worst in five years of doing business with the organization, and they were initiating a contract review.

What happened? The aggregate metrics were accurate. They were also misleading. The customer satisfaction improvement was driven entirely by a spike in favorable scores among new, low-spend customers who had experienced a recently improved onboarding flow. The legacy client segment — older, larger accounts with more complex service needs — had experienced a measurable decline in satisfaction that was entirely invisible in the blended average. The first-call resolution improvement reflected a change in how cases were categorized, not a change in actual resolution quality. The handle time reduction reflected an automation change that resolved simple cases faster, while routing complex cases to a callback queue with longer wait times. Every metric told a true story about a subset of reality. The aggregate story was demonstrably false for the accounts that mattered most.

Data-driven decision-making does not mean trusting dashboards. It means understanding what data reveals, what it conceals, and where the boundary between the two lies. That understanding is a management skill — not a technical one, not a data science credential — and it is the skill this chapter is designed to build. Managers who develop genuine data literacy do not need to run their own statistical analyses. They need to know which questions to ask, which metrics to trust and which to interrogate, how to recognize when a number is accurate but misleading, and how to communicate evidence-based decisions to stakeholders who may hold strong prior views. These are leadership capabilities, and they are learnable by any manager willing to think rigorously about the evidence behind their recommendations.

Diagram 12.1: The Misleading Dashboard Scenario

Surface data

Segmented data

Leadership

SALES GROWTH +13.1%

CUSTOMER ACQUISITION +2.40/1.0%

OVERALL REVENUE +1.3%

key segment

+5   +2   +3   -1.0   -6   -5.4 -6.2

Simplified Dashboard | Manager | Segmented Data Breakdown

Aggregate dashboards can obscure significant declines in critical segments, requiring granular analysis for accurate business insights.

13.2   Why Data Literacy Is the Modern Manager's Decision Advantage

The volume of data available to managers has increased by an order of magnitude over the past decade, and the trend is accelerating. Operations platforms, customer relationship systems, HR tools, financial reporting systems, project management platforms, and AI analytics tools all produce data continuously. A manager who cannot navigate this data environment effectively is not working with less information than their peers who can — they are working with more information that they are less equipped to use. The result is decisions that are less accurate, less defensible, and less aligned with the operational reality they are meant to address.

The decision quality advantage of data literacy is well-documented and consistent. Research across industries shows that managers who incorporate structured data analysis into their decision-making process make measurably better decisions on operational questions — resource allocation, project prioritization, risk assessment, and performance evaluation — than those who rely primarily on intuition and experience. This does not mean intuition is worthless: experienced managers bring pattern recognition and contextual knowledge that data cannot fully capture. It means that intuition without data validation is vulnerable to a class of systematic errors — availability bias, anchoring, overconfidence — that structured data examination can correct.

The accountability advantage of data literacy is equally significant in organizational contexts. When a manager makes a decision challenged by leadership, peers, or direct reports who disagree, the ability to articulate the evidence base for that decision is a professional asset that data-literate managers possess and that data-averse managers do not. 'I decided based on these specific data points, which showed this pattern, which my analysis suggested was caused by these factors' is a more defensible and more credible position than 'I decided based on my read of the situation.' The former invites productive engagement with the evidence; the latter invites subjective disagreement. In both performance conversations and organizational influence, data literacy creates professional leverage.

AI tools now multiply the value of data literacy by making sophisticated analysis more accessible. AI-powered analytics platforms can surface patterns in large datasets that would otherwise require significant manual analysis to identify, generate predictive scenarios for planning decisions, and produce natural language summaries of complex data, making analysis results accessible to non-technical stakeholders. But these tools amplify the manager's data literacy rather than substituting for it. A manager who does not understand the basics of data interpretation will misread AI-generated analysis as reliably as they misread manual analysis — perhaps more so, because AI-generated outputs often carry a veneer of authority that suppresses the skepticism they deserve. Building your own data literacy is a prerequisite to getting value from AI analytics tools, not an alternative to using them.

The cultural dimension completes the picture. A manager who makes data-visible decisions — who shows their team the evidence behind choices, who asks for data before committing to directions, who changes their position when evidence warrants it — builds a team culture that is more intellectually honest, more effective at self-correction, and more resilient in navigating uncertainty. Teams where decisions are made based on relationship dynamics, political influence, and who speaks most confidently rather than on evidence have more volatile performance patterns and more persistent blind spots. The manager's data behavior sets the standard for the team's data culture, and that culture compounds over time into organizational capability.

Manager to have three decision advantages of manager data literacy, a concise advantage in success.

13.3   The Decision-Making Spectrum: Intuition, Data, and AI-Assisted Judgment

The framing of 'data-driven' versus 'intuition-based' decision making is a false dichotomy that creates as many problems as it solves. Real management decisions exist on a spectrum, and the appropriate position on that spectrum depends on the nature of the decision, the quality and relevance of available data, the stakes and reversibility of the choice, and the time available for analysis. Dogmatic data-drivenness produces analysis paralysis, misleads when data quality is poor, and alienates team members who understand that some questions cannot be reduced to numbers. Dogmatic intuition-based deciding produces inconsistency, perpetuates bias, and fails on complex, multi-variable problems where human pattern recognition is systematically unreliable. The goal is calibrated judgment — knowing where on the spectrum a given decision belongs and why.

At the intuition dominant end of the spectrum sit time-critical decisions that involve novel situations without historical precedent, require reading interpersonal dynamics not captured in data, or depend on values and priorities that are not reducible to metrics. An immediate escalation decision during a client crisis. A judgment call about how a specific team member will respond to a particular feedback approach. A reading of organizational political dynamics that affects how to sequence a change initiative. For these decisions, data may inform the context. Still, the decision itself depends on the manager's judgment, experience, and situational awareness in ways that cannot be fully specified in advance or validated by a dataset.

At the data-dominant end of the spectrum sit decisions that are recurring, high-volume, or consequential enough to warrant structured analysis — and for which quality data is available and relevant. Resource allocation across a portfolio of workstreams. Assessment of which product features are driving customer retention. Evaluation of vendor performance against defined service level agreements. Identification of leading indicators that predict project schedule risk. For these decisions, intuition without data validation is vulnerable to systematic errors that data examination can correct. The manager's role is to ensure the analysis is correctly framed, the data quality is adequate, and the conclusions drawn from the data are logically valid — not to conduct the analysis personally, but to be able to evaluate it critically.

The middle of the spectrum — where most consequential management decisions actually live —

combines both. Strategic resource allocation decisions involve data on past performance and projected demand, along with the manager's judgment about team capabilities, development goals, and organizational dynamics that are not fully captured in the data. Performance improvement decisions involve quantitative performance data and qualitative assessment of motivation, context, and development potential. Risk management decisions involve quantitative risk modeling and qualitative assessment of organizational risk tolerance and strategic priorities. For these decisions, the manager's value-add is the synthesis: using data to discipline intuition, and using judgment to interpret data in ways that account for context the data cannot capture.

Diagram 12.3: The Decision-Making Spectrum: Intuition to Data to AI

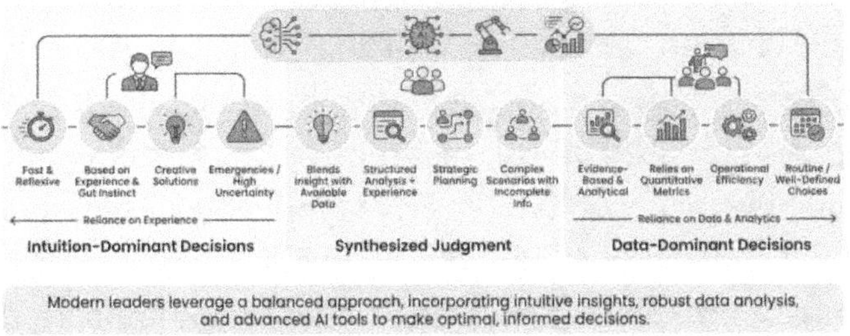

Modern leaders leverage a balanced approach, incorporating intuitive insights, robust data analysis, and advanced AI tools to make optimal, informed decisions.

13.4   Building a Data-Informed Culture on Your Team

Culture does not change by decree. A manager who announces that the team will henceforth make all decisions based on data will find, six months later, that the team is doing exactly what it was doing before, and that the team has added a data slide to the end of every presentation to comply with the announcement. Culture changes through consistent behavior — through the decisions the manager makes visibly and how they make them, through what questions get asked in every team meeting, through what gets celebrated and what gets corrected, and through the norms and expectations that shape everyday work. Building a genuinely data-informed team culture requires deliberate and sustained behavioral modeling by the manager, not a one-time training initiative.

13.4.1 The Data Question Habit

The single most powerful culture-shaping behavior a manager can adopt is consistently asking for data before endorsing a recommendation. When a team member presents a proposal, ask: What does the data show about this question? When a problem is described in a team meeting, ask: What do we know about the scale and pattern of this problem, and where is that knowledge documented? When a success is reported, ask: how do we know this worked, and what evidence would tell us it is replicable? These questions do not communicate distrust — when framed with genuine curiosity rather than challenge, they communicate that the manager values evidence and expects it as a standard part of how the team thinks. Over time, that expectation reshapes how the team prepares for every conversation with the manager and among themselves.

The complementary behavior is to update your position when the data warrants it visibly. When evidence challenges a prior view, say so explicitly: 'I thought this was the case, but this data suggests I was wrong, and I am updating my thinking accordingly.' That behavior models epistemic honesty in a way that gives team members explicit permission to do the same — to change their minds based on evidence without experiencing it as an admission of incompetence. Teams in which position changes in response to evidence are valued produce better decisions than teams in which consistency of view is valued, because the former teams surface and correct errors earlier and with less political friction.

13.4.2 Shared Metrics and Transparent Performance Visibility

Data-informed culture requires shared access to data describing team performance, not just top-down reporting of results. When team members can see the same performance metrics the manager sees — project velocity, quality indicators, customer satisfaction trends, utilization patterns, risk flags — they can make better day-to-day decisions without requiring managerial approval at every step, because they have the context to understand how their individual choices connect to team-level outcomes. This visibility also creates natural accountability that does not depend on surveillance. When team members see that their work decisions are reflected in shared performance data, they develop a more direct relationship with the outcomes they are producing.

The design of shared performance dashboards is a leadership decision, not just a technical one. Which metrics to make visible, how to present them in ways that are interpretable without extensive training, how to prevent metric gaming where team members optimize measured indicators rather than actual outcomes — these choices reflect the manager's theory of what good performance looks like and how it should be measured. Involve the team in dashboard design and metric selection where possible: teams that have participated in defining their own performance measures are more likely to trust those measures and less likely to view them as external control mechanisms.

Tit. diagram 12.4: Building data culture: manager behaviors that shape team norms.

13.5   Dashboards, Metrics, and KPIs That Actually Matter

The proliferation of data tools has made it easier than ever to measure almost anything — and that abundance poses a danger rarely discussed: measuring the wrong things with increasing precision. A team drowning in metrics is not a data-informed team. It is a team spending cognitive resources on measurement rather than on the work the measurement is supposed to improve. The discipline of metric selection — choosing the fewest indicators that together provide an accurate, actionable picture of team performance — is one of the most practically valuable data skills a manager can develop.

A workable framework for metric selection distinguishes between leading indicators, lagging indicators, and diagnostic metrics. Leading indicators are forward-looking measures that predict future outcomes: the current pipeline of work, early quality signals in process, team capability development progress, and risk flag counts in project monitoring. They are valuable because they create opportunities for intervention before problems manifest in outcomes. Lagging indicators are backward-looking measures that confirm whether past performance produced the desired results: project delivery rate against plan, customer satisfaction scores, revenue, or cost performance. They are essential for accountability and learning, but provide no opportunity for in cycle correction. Diagnostic metrics provide context for interpreting both leading and lagging indicators: utilization rates, team composition stability, external environment factors, and process adherence rates.

The practical standard for a team KPI set is that every metric should pass three tests. First, it must be actionable: if the metric moves in the wrong direction, there must be a specific set of management actions that could address it, and those actions must be within the manager's authority or achievable through clear escalation. Metrics that cannot be acted upon are noise, not signal. Second, it must be attributable: the metric should primarily reflect the team's own performance rather than external factors beyond the team's control. A metric that is seventy percent driven by market conditions tells you more about the market than about the team. Third, it must be resistant to gaming. If optimizing the measured number does not require actually improving the underlying performance, the metric creates perverse incentives that will be discovered and exploited. Metric design is a governance exercise as much as a measurement exercise.

Apply the same rigor to AI-generated metrics and KPIs. AI analytics tools can automatically surface new metrics — patterns in data that human analysis would not have identified — and this can be genuinely valuable when it reveals leading indicators of problems that would otherwise remain invisible. But AI-generated metrics should pass the same three tests as manually designed ones: actionable, attributable, and gaming-resistant. An AI tool that surfaces a metric you do not understand well enough to evaluate on these dimensions requires investigation before adoption, not automatic trust because it was algorithmically generated.

Diagram 12.5: KPI Selection Framework: The Three-Test Filter

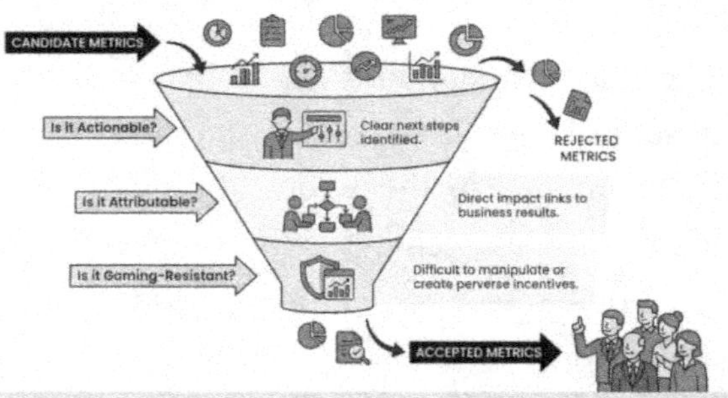

Using a rigorous selection process ensures only the most meaningful Key Performance Indicators drive business strategy.

13.6 Using AI for Predictive Analytics and Scenario Planning

Predictive analytics — using historical data and statistical models to forecast future outcomes — has been transformed by AI tools from a capability requiring dedicated data science resources into one increasingly accessible to managers without specialist technical expertise. The tools available today can generate demand forecasts, project schedule risk predictions, attrition risk scores, and revenue scenario models with substantially less setup than their predecessors required. For managers, this creates a genuine decision-support advantage — but only when used with an accurate understanding of what prediction means and its limits in practice.

Predictive analytics outputs are probabilistic, not deterministic. A model that projects a 65% probability of a project delivery delay does not predict that the project will be late. It is saying that projects with these characteristics have historically been late sixty-five percent of the time — and that historical pattern may or may not apply to your specific project, depending on how similar your current context is to the training history. The practical value of the prediction is not certainty about the outcome; it is identifying which workstreams or situations warrant proactive attention and intervention. The manager's job when receiving a predictive output is to ask: what specific features are driving this prediction, are those features actually present in the current situation as the model assumes, and, if the prediction

338

is accurate, what actions would I take differently starting today?

Scenario planning extends predictive analytics from single-point forecasts to structured exploration of alternative future states. Rather than asking 'what will happen?' — a question no model can answer with certainty — scenario planning asks 'what would happen under each of these plausible sets of conditions, and how should we prepare for each?' AI tools make scenario planning more accessible by dramatically reducing the computational and analytical overhead of modeling multiple scenarios. A manager can define two or three key uncertainty dimensions — demand level, resource availability, technology timeline — and an AI planning tool can generate outcome projections for the combination of states along those dimensions, giving the manager a structured picture of the decision space rather than a single projected path.

The governance discipline for predictive analytics and scenario planning is scenario validation. Before relying on a predictive model's output for a significant planning decision, the model's assumptions should be made explicit and reviewed against current reality. What historical period does the training data cover? What were the primary drivers of the predicted outcomes in that data? Are those drivers still operating in the same way in the current environment? Scenario validation does not require rebuilding the model — it requires the manager to ask the team or vendor that manages the model those questions and evaluate whether the answers are consistent with what they know about the current operating context.

The most valuable discipline in scenario planning is not the quality of the scenarios themselves but the pre-commitment to decision triggers: defining in advance what observable conditions would cause you to move from one planned response to another. Pre-committed decision triggers eliminate the delay and ambiguity of real-time decision-making under uncertainty, reduce the influence of in the moment cognitive biases, and create accountability for following through on the analysis work that produced the scenarios. When the trigger condition is met, the response is already decided — not improvised under pressure.

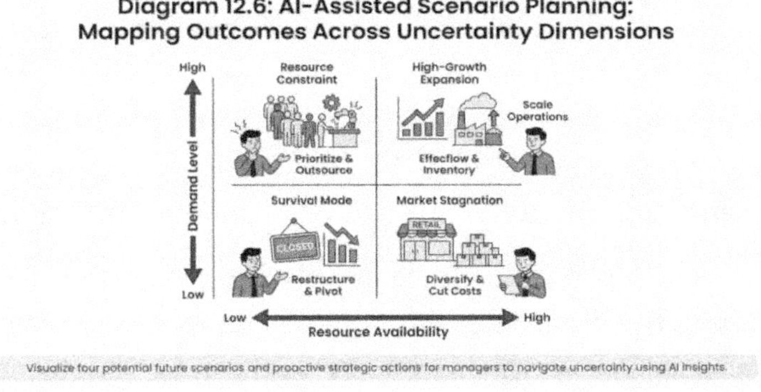

13.7   Avoiding Common Data Pitfalls

Data literacy is not just knowing how to use data effectively — it is knowing how to avoid being misled by it. The most consequential data errors in management contexts are not technical calculation errors, which modern tools catch automatically, but interpretive errors: conclusions drawn from data that the data does not actually support. These errors are systematic and predictable, which means they can be specifically guarded against once a manager knows what to look for.

13.7.1 Correlation and Causation

The correlation/causation error is the most frequently committed analytical mistake in business settings, and it is more seductive when the correlation is produced by an AI tool than when it is observed manually, because the algorithmic generation of the pattern lends it an air of objectivity it does not deserve. The pattern is this: two variables that move together over time do not necessarily have any causal relationship. Teams that use more AI tools and those with higher performance may both hold for the same set of high-resource, well-led organizations — but that correlation does not mean AI tool adoption causes better performance; both may be the result of the underlying quality of team leadership and organizational investment.

The test for whether a correlation can support a causal claim requires asking three questions: Is there a plausible mechanism by which changes in the first variable would produce changes in the second? Is there a logical time sequence — does the proposed cause precede the proposed effect? And has the relationship been tested after controlling for other variables that might explain both? These questions do not require statistical sophistication to apply. They require intellectual discipline and the willingness to challenge an attractive conclusion. When an AI analytics tool surfaces a correlation that supports a decision you were already inclined to make, that is precisely the moment to apply the most rigorous scrutiny — confirmation of prior beliefs is the environment where correlation/causation errors are most commonly made.

13.7.2 Confirmation Bias and Selective Data Use

Confirmation bias — the tendency to seek, interpret, and remember information in ways that confirm prior beliefs — is one of the most robust and replicable findings in behavioral psychology, and it operates in data analysis as reliably as in every other domain of human judgment. Managers who have already formed a view about a question tend to find the data that supports that view and discount the data that challenges it, often without conscious awareness that they are doing so. AI tools do not eliminate this bias; they can amplify it if they are used to search for supporting evidence rather than to test a hypothesis against all relevant evidence.

The practical safeguard is a structured analysis discipline: define the question, define what evidence would support the null hypothesis as well as the hypothesis you are testing, look for both, and document the analysis process before concluding. This pre-commitment to symmetrical evidence evaluation is harder than it sounds in a real organizational environment where there is often time pressure, stakeholder pressure to reach a particular conclusion, and the natural human preference for coherence and resolution. But the managers who develop this discipline make fewer consequential errors than those who do not, and over time their credibility as decision-makers reflects that accuracy.

13.7.3 The Denominator Problem and Base Rate Neglect

A common data presentation error in management reporting is the selective emphasis of numerators without adequate context from denominators. The number of customer complaints received this month means something very different when compared against 10,000 transactions than against 1,000 transactions. The number of projects completed on time is meaningful only in the context of the total projects in the portfolio. The number of employees who raised a particular concern in a survey means something different if it represents 3% of respondents than if it represents 40%. Always ask: what is this number a fraction of, and is the denominator comparable across the periods or groups I am comparing?

Base rate neglect is a related error: overweighting vivid, specific case information relative to the broader statistical base rate. When an AI attrition model flags three team members as high flight risk, the natural response is to focus intensively on those three individuals. But the relevant context question is: what is the model's base rate accuracy — how often do employees flagged as high risk actually leave, and how often do employees not flagged leave unexpectedly? A model that is right 60% of the time is useful; a model that is right 40% of the time wastes intervention resources and creates awkward relationships with team members who were wrongly profiled. Knowing the base-rate accuracy of any predictive tool you use is essential for calibrating how much weight to place on its outputs.

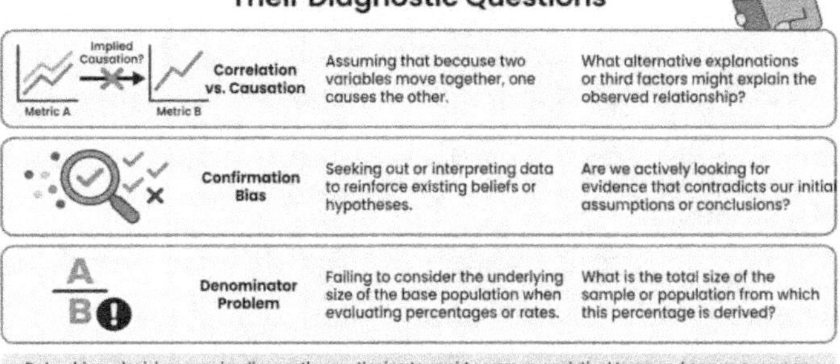

13.8 Communicating Data-Driven Decisions to Stakeholders

Making a data-informed decision is necessary but not sufficient. Communicating that decision in a way that stakeholders can understand, trust, and act on is the skill that translates analytical rigor into organizational impact. Managers who make excellent evidence-based decisions but communicate them poorly — either drowning stakeholders in data or presenting conclusions without evidence — lose the organizational benefit of the analysis. The goal of communicating a data-informed decision is to give stakeholders enough context to understand what the evidence showed, why you drew the conclusions you did, and what you are asking them to do or approve — without requiring them to verify the analysis independently.

The most effective structure for communicating a data-informed decision is a three-layer narrative that works from conclusion to evidence to implications, not from data to conclusion to request. Start with the bottom line: what did you decide, and what is the most important single fact that supports that decision? Stakeholders are busy and context-dependent; they need the conclusion and its primary support before they can engage productively with the nuance. Then provide the evidence layer: what data did you examine, what pattern did it reveal, and how confident are you in the interpretation? This layer should be honest about uncertainty — 'this data strongly supports the direction' and 'this is a reasonable inference from limited data but carries higher uncertainty' are both legitimate framings that build more long-term credibility than false confidence. Finally, provide the implications layer: what specific actions follow from the decision, what the stakeholder needs to do or provide, and what signals you will monitor to assess whether the decision is producing the intended results?

Visual data presentation requires its own discipline. The most common error in stakeholder data presentations is showing too many charts without explaining what each one means. A slide with five charts and no narrative interpretation forces stakeholders to draw their own conclusions — which may not match yours, especially if they are encountering the data for the first time without the analytical context you developed over days of examination. The principle of one chart per insight — where each visualization is accompanied by a clear statement of what it shows and why it matters — yields faster comprehension and more productive engagement than data dumps that require the audience to do the presenter's analytical work for them.

Handling disagreement with data-backed decisions requires a specific approach that many managers find uncomfortable: being willing to say 'I understand your intuition, and I had a similar one before I saw this data — let me show you what changed my thinking.' This framing validates the stakeholder's perspective, demonstrates intellectual honesty about your own prior view, and invites engagement with the evidence rather than creating a positional standoff. Stakeholders who feel that their instincts were heard before being challenged by evidence are significantly more likely to update their views than those who experience data presentation as dismissal of their judgment.

Diagram 12.8: Communicating Data-Driven Decisions: The Three-Layer Structure

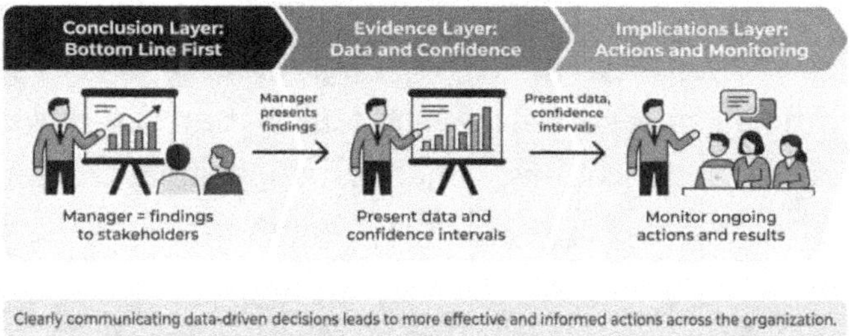

Clearly communicating data-driven decisions leads to more effective and informed actions across the organization.

13.9   Balancing Speed and Rigor in Decision-Making

The organizational pressure on managers to decide quickly is real and often appropriate — markets move, customers do not wait, operational windows close. But speed and rigor are not inherently in opposition. The manager who has built data literacy, established clear metrics, and created accessible performance visibility can make faster data-informed decisions than the manager who must gather and interpret data from scratch for every decision. Data infrastructure — the combination of tools, practices, and governance that makes relevant data readily accessible — is what makes speed and rigor compatible rather than competing values.

The decisional framework that best balances speed and rigor distinguishes between decisions by their reversibility and stakes rather than by available time alone. Highly reversible decisions with limited stakes — which communication channel to use for a team announcement, how to sequence tasks within a sprint, whether to hold a meeting synchronously or asynchronously — should be made quickly, often without extensive data analysis, because the cost of a wrong decision is low and the cost of analytical delay is non-trivial. Highly irreversible decisions with significant stakes — hiring and termination decisions, major resource reallocations, changes to team structure and mission, technology platform commitments — warrant analytical rigor proportional to their consequences, even when organizational pressure suggests speed.

The practical discipline for time-pressured high-stakes decisions is satisficing with defined criteria rather than optimizing with open-ended search. Before beginning the analysis, define the minimum evidence standard that a decision alternative must meet to be acceptable — specific performance thresholds, confidence level requirements, risk limits. Then analyze until you have found an alternative that meets those criteria, not until you have analyzed every conceivable option. This approach produces decisions that are demonstrably evidence-based and defensible without falling into analysis paralysis. The criteria definition is the analytical discipline; the search stops when a criterion-meeting option is found, not when time runs out, or exhaustion sets in.

AI-assisted analysis tools improve the speed-rigor balance by dramatically reducing the mechanical work of data gathering, processing, and initial pattern identification — work that previously required hours of analyst time now requires minutes of AI processing. The manager's contribution is the framing, the contextual interpretation, the criterion definition, and the final judgment call that integrates quantitative signals with qualitative context. When AI does the mechanical work and the judgment work is done by the manager, decisions can be both faster and more rigorously evidence-based than either pure intuition or purely human-executed analysis alone can achieve. That combination — AI-accelerated data processing and human-exercised judgment — is the decision-making model that characterizes the most effective managers in the current era.

13.10 Manager's Checklist: Data-Driven Decision Making in Practice

• Conduct a metric audit for your team: eliminate any KPI that fails the actionable, attributable, or gaming-resistant test, and replace it with an indicator that passes all three.

• Establish a practice of defining the minimum evidence standard for major decisions before beginning the analysis — document the criteria and share them with stakeholders before the analysis starts.

• Review your team's performance dashboard with a fresh eye: does each metric tell you something actionable, or is it there because it was always there? Remove noise ruthlessly.

• Build the data question habit: in every team meeting where a recommendation is made, ask what data supports the recommendation before endorsing or advancing it.

• Identify the top five data sources your team uses most frequently and assess each for recency, representativeness, and known limitations — document those limitations in a shared team reference.

• For every predictive analytics tool your team uses, document in writing what historical period the model was trained on, what the primary outcome predictors are, and what the base rate accuracy is.

• Practice the three-layer communication structure — conclusion, evidence, implications — for every data-informed decision you present to leadership or cross-functional stakeholders.

• Review the last three significant team decisions: for each, identify whether the final choice was supported by data, whether alternative explanations for the data were considered, and whether the correlation/causation discipline was applied.

• Create a shared team space — even a simple shared document — where AI-generated analytics outputs are documented along with the human interpretation and decision action that followed, building an institutional record of how data was used.

- Schedule a quarterly data retrospective with your team: review which metrics moved, what decisions followed, whether outcomes matched expectations, and what the data tells you about which analytical approaches are working and which need revision.

13.11 What You Now Carry Forward

The dashboard that misled your leadership team was not due to bad data. It was the result of a gap between what the data measured and what the organization needed to know — a gap that data literacy closes and that data absence or data avoidance leaves open and dangerous. The customer who called to initiate a contract review was not a data problem. They were a measurement design problem, a segmentation problem, and an interpretive problem — all of which are management problems that the principles in this chapter directly address.

Data-driven decision making is not about trusting numbers over people, or about eliminating the human judgment that is irreplaceable in every consequential management decision. It is about disciplining that judgment with evidence, testing intuitions against data before acting on them at scale, building the metrics infrastructure that gives you and your team an accurate and actionable picture of performance, and communicating evidence-based decisions in ways that build stakeholder trust and organizational credibility. These are leadership practices that make you more effective, not technical practices that require you to become a data scientist.

The AI tools that are now available to support data-informed management — predictive analytics, natural language data querying, automated scenario modeling, AI-generated insights and summaries — amplify the value of data literacy rather than substituting for it. The manager who brings rigorous data thinking to AI-generated outputs will use those tools effectively, catch their errors, and make better decisions with them than without. The manager who defers to AI outputs without data literacy will replicate the original dashboard's problem: receiving technically accurate information that is misleading in context. Your data literacy is the safeguard — for AI tools as for all analytical instruments — and it is the foundation of the operational clarity that every effective manager works toward.

The decisions ahead of you — about resources, people, strategy, technology, and risk — deserve that rigor. Your team deserves a manager who makes those decisions with both the intellectual honesty to seek disconfirming evidence and the courage to act on what the evidence actually shows, even when it contradicts a prior view or a politically convenient narrative. That combination of rigor and courage is what data-driven leadership means in practice. The tools and frameworks in this chapter give you the foundation. The practice — asking the questions, building the habits, modeling the discipline — is what you bring to it.

14 Leading Through Change and Digital Transformation

14.1 Scenario

Three months ago, your organization announced a company-wide shift to a new AI-powered workflow platform. The executive team called it a "transformational leap forward." Your team called it something else. Two of your strongest performers submitted informal requests to transfer to other departments. A third team member — someone you depend on for institutional knowledge — told you privately that she planned to retire early rather than "start over learning machines." Meanwhile, your manager expected you to hit the same productivity targets during the transition and report green on a dashboard that had no column for "team morale at an all-time low."

You didn't cause this change. You didn't design it, fund it, or schedule it. But you are the person your team looks to when the uncertainty becomes unbearable, and you are the person your leadership looks to when results are at risk. That is the manager's position in every organizational transformation: responsible for outcomes you did not choose, for people experiencing emotions you cannot always predict, during timelines set by people who will not feel the turbulence the same way your team will.

This chapter equips you to navigate that position with clarity, confidence, and operational precision. It is not a chapter about surviving change. It is a chapter about leading it — actively, deliberately, and in a way that leaves your team more capable at the end than they were at the start. Every manager reading this has been through some version of this scenario, whether the change involved new technology, organizational restructuring, a leadership transition, or a market shift that invalidated established ways of working. You are not the first to stand in this position, and the body of management knowledge that exists to support you in it is more useful than most transformation programs let on.

Diagram 13.1: The Manager's Position in Organizational Change

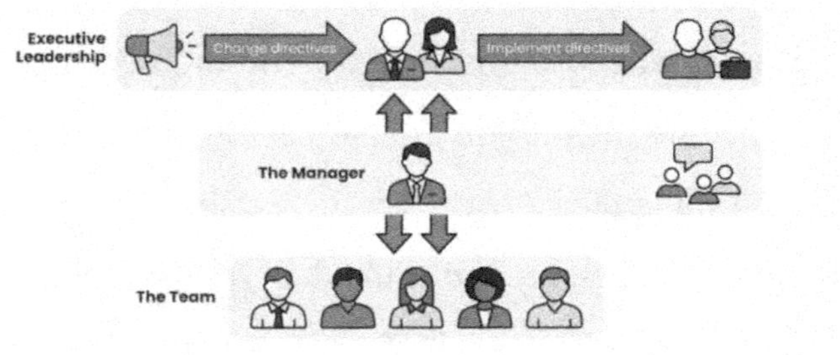

The manager wilfte genels ts translator, ulT between corporate strategy and team execution.

14.2 **Why It Matters**

Change is no longer an event that happens between two stable periods. For most organizations, continuous change is the operating condition, and the AI transformation currently reshaping every industry has accelerated the pace. Teams that struggled with major system upgrades every three or four years now face iterative tool changes, process redesigns, and evolving capability expectations that evolve quarter to quarter. This is not a temporary disruption — it is the new normal, and the competency required to lead through it is now a baseline expectation for every manager, regardless of industry or function.

The cost of managing change poorly is not abstract. Research across industries consistently shows that the majority of organizational change initiatives fail to deliver their intended results, and the most common failure point is the middle layer of management. Executives sponsor change; frontline employees experience it, but managers are the mechanism through which change either takes root or collapses. When managers lack frameworks, language, and authority to lead change competently, change becomes something that happens to their teams rather than something they move through together. The productivity loss during poorly managed transitions is compounded by the talent loss that follows — top performers leave when their managers cannot provide credible navigation.

For new managers in particular, leading through transformation is a defining test. The team is watching closely to see whether their manager is a genuine source of stability or simply another source of anxiety in uniform. The organization is watching to see whether this manager can handle complexity. And the manager is watching herself, often for the first time in a situation where neither technical expertise nor individual performance can solve the problem at hand. The skills required are relational, communicative, and adaptive — the full portfolio of what leadership actually demands.

The AI dimension adds a layer of complexity that previous generations of managers did not face. When change involves AI tools, it triggers fears that go beyond workflow disruption: fears about relevance, about professional identity, about whether the skills that made someone valuable are about to become obsolete. A manager who treats AI adoption as a simple technical training problem will lose team members not to other companies but to discouragement. Addressing the human dimension of AI-era change is not a soft skill add-on — it is mission-critical. The organizations that successfully integrate AI into their workflows are those in which managers made the human architecture of adoption as rigorous as the technical architecture.

Diagram 13.2: Why Change Initiatives Fail: The Manager Layer

POOR COMMUNICATION

MANAGER DISENGAGEMENT

UNCLEAR EXPECTATIONS

INADEQUATE TRAINING

RESISTANCE NOT ADDRESSED

Mixed messages

Ignoring

Vaguely

Watching

Effective managers are critical to successful change; neglecting their engagement and capability often leads to initiative failure.

14.3   Change Management Fundamentals

Every manager leading through transformation needs a working knowledge of the established frameworks that describe how change unfolds and how people move through it. You do not need to memorize academic models. Still, you do need to understand the underlying logic so you can diagnose where your team is at any given moment and respond accordingly. The value of these frameworks is not that they predict what will happen — no model does that reliably — but that they provide a vocabulary for what is already happening and a set of intervention points you would otherwise miss.

14.3.1 Kotter's Eight-Step Model

John Kotter's eight-step model, developed over decades of studying organizational transformation, describes change as a sequence: create urgency, build a guiding coalition, form a strategic vision, enlist a volunteer army, enable action by removing barriers, generate short-term wins, sustain acceleration, and institute change. While Kotter's original model was designed for large-scale organizational transformation, managers can draw on its logic at the team level. The steps do not have to be sequential at the team scale — you can work on several simultaneously — but the underlying logic of each is sound.

The most important insight from Kotter is that urgency is real motivation, not manufactured pressure. Managers who announce change by saying "we have no choice" are invoking urgency without creating it. Real urgency comes from helping your team understand what is at stake: for the organization, for the team, and for each individual. When people understand the genuine cost of not changing, they become participants in the transition rather than passengers. The urgency conversation is most effective when it is specific — not "the market is demanding AI capabilities" but "here is how our current workflow creates a competitive disadvantage that affects our team's workload and our department's standing."

The concept of short-term wins is equally valuable for team-level change management. When a major transformation is announced, the destination feels remote, and the path feels treacherous. Short-term wins are checkpoints along the path that demonstrate progress, build confidence, and give the team evidence that the effort is paying off. As a manager, engineering these early wins is one of your most powerful levers. Identify the first capability your team can demonstrate with a new tool, celebrate it visibly, and use that moment to make it feel achievable. A team that has experienced one genuine win with a new system approaches the next challenge with measurably more confidence than a team that has only been told the system will be beneficial.

14.3.2 The ADKAR Model

The ADKAR model, developed by Prosci, takes an individual centered view of change. Rather than describing organizational stages, it describes the psychological journey each person must complete to fully adopt a change:

Awareness of the need for change, Desire to support and participate, Knowledge of how to change, Ability to demonstrate skills and behaviors, and Reinforcement to sustain the change. The power of ADKAR is in its precision — it does not treat change adoption as a uniform process but as a sequence of specific milestones that can be assessed and supported individually.

ADKAR is particularly useful for managers because it reveals that people on the same team can be at different stages simultaneously. One team member may have full awareness and a strong desire, but lack knowledge — they want to use the new AI platform but have not had sufficient training. Another may have knowledge and ability, but weak desire — they understand the tool but have not yet committed to changing their habits. Managing change effectively means diagnosing where each person is and applying the right support rather than running the same change management program for everyone. A training session helps someone in the Knowledge gap; it does not move someone who is stuck in the Desire gap.

When you work through the ADKAR lens with your team, a checklist conversation is more useful than a presentation. Ask each team member directly: "Do you understand why we're making this change?" (Awareness) "Do you want to make this work?" (Desire) "Do you have the training and information you need?" (Knowledge) "Do you feel confident doing this in practice?" (Ability) "Is there enough support in place to keep this going?" (Reinforcement). Their answers tell you where to focus your energy. A manager who asks these questions and then provides targeted, individualized support will consistently outperform one running a generic change management program.

Diagram 13.3: ADKAR Individual Change Journey

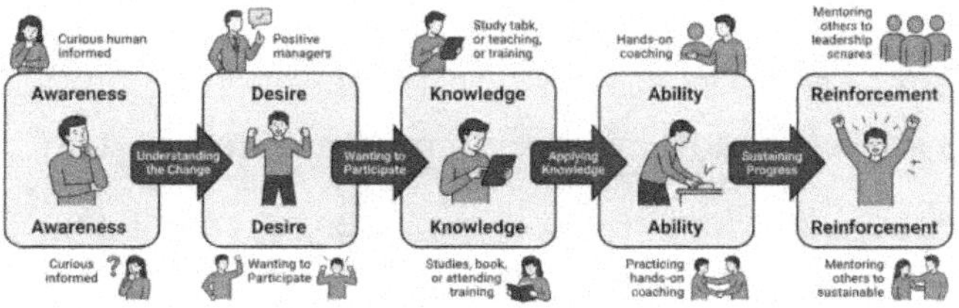

ADKAR stages is aratting change model. Itatiin each person change through, individual chaney.

14.3.3 Managing the Emotional Arc of Change

Neither Kotter nor ADKAR fully captures the emotional experience of change — the grief, frustration, fear, and eventual acceptance that characterize how people actually feel during transformation. The change curve, adapted from Elisabeth Kübler-Ross's model of the stages of grief, maps this emotional terrain: shock, denial, frustration, depression, acceptance, decision, integration. Not everyone moves through these stages in sequence, and not everyone reaches integration on the same timeline. But the pattern is consistent enough to be useful as a diagnostic tool.

As a manager, you do not need to be a therapist, but you do need to be literate in this emotional language. When a team member snaps at you in a meeting about the new system, that is not insubordination — that is frustration. When another goes silent and stops engaging, that is not disengagement — that may be depression in the change curve sense, a retreat from participation that signals someone has not yet found a reason to commit. Responding to these signals with administrative pressure rather than human attention accelerates the very disengagement you need to prevent. Pressure applied to someone in the frustration or depression stage does not produce acceleration — it produces either compliant non-adoption or departure.

The manager's job in the emotional arc of change is to normalize the experience without validating dysfunction. Saying "it makes sense that this feels disruptive — let's talk about what specifically is hardest for you" is different from saying "yeah, this is terrible, I don't know why they're making us do this." The first acknowledges the human experience and opens a productive conversation. The second aligns you with resistance and undermines your ability to lead. Normalization without alignment with resistance is a finely calibrated posture — and it separates managers who create genuine safety during transformation from those who inadvertently deepen it.

14.4 The Manager as Change Agent and Translator

There is a gap between the strategy and the team. Executives think in quarterly outcomes, capability investments, and competitive positioning. Frontline employees think in daily tasks, familiar routines, and the specific skills they have spent years developing. Neither group is wrong. They are operating at different levels of abstraction. The manager's role is translation — not neutral relay but active conversion of meaning across two different operating languages.

Translation in this context means three things. First, it means converting the organization's change rationale into language that is meaningful at the team level — not "we are accelerating our AI transformation to remain competitive," but "the way we build reports is going to change, and here is what that means for your day-to-day workflow and how I plan to support you through it." Second, it means carrying your team's ground-level experience back up to leadership — surfacing the real barriers, the actual gaps in training, the unintended consequences that are not visible from the executive layer. Third, it means holding both layers accountable: pushing back on unrealistic timelines, requesting resources the team genuinely needs, and refusing to relay executive communication without adding your own honest interpretation.

This translation function requires courage. It is easier to pass information up and down without engaging critically at either level. But passive translation is not management — it is coordination. The manager who adds value in a transformation is the one who stands in the middle of the tension and does not flinch. When the executive layer sends a message that is technically accurate but emotionally tone-deaf, the manager's job is to translate it into language the team can actually receive. When the team is expressing concerns that leadership has not heard, the manager's job is to carry those concerns upward with the force and specificity they require.

14.4.1 Building Your Change Communication Cadence

Communication during change cannot be left to organizational announcements and quarterly all-hands meetings. Transformation demands a local communication cadence — one owned by you, tailored to your team, and operating at a frequency appropriate to the level of uncertainty. The higher the uncertainty, the higher the required communication frequency. During the peak of a major transition, daily brief updates may be more appropriate than a weekly comprehensive summary.

The fundamental principle is this: when information is absent, people fill the gap with their worst assumptions. A team that hears nothing from its manager about a major change in progress will generate its own narrative. That narrative is almost always darker and more catastrophic than reality. Your job is to occupy that narrative space with accurate, honest, regular information — even when the information is "we don't know yet." The admission of uncertainty, delivered consistently, builds more trust than confident predictions that turn out to be wrong.

A practical communication cadence during transformation includes a weekly team check-in with an explicit change update segment, brief individual check-ins with team members who are showing signs of stress or disengagement, a dedicated channel or thread for questions about the change where you respond promptly, and a standing escalation path to your manager for questions you cannot answer. The frequency and format can adapt over time, but the consistency cannot. Teams in transition need to know they can count on a regular, honest communication channel — and that channel is you.

Be direct about what you know and what you do not know. "I don't have an answer to that yet, but I'll find out and get back to you by Thursday" is significantly more credible than a vague reassurance that everything will work out. Team members notice when you answer every question with confidence you don't actually have — and they lose trust in direct proportion to those failures. The manager who builds a track record of honest uncertainty acknowledgment and consistent follow-through earns a level of credibility that carries the team through moments when the honest answer is genuinely unknown.

14.5 Leading AI Adoption: Overcoming Fear, Building Enthusiasm

AI adoption occupies a particular category of organizational change because it triggers identity-level fears. When a new scheduling system is introduced, people worry about learning a new interface. When AI tools are introduced — especially those that automate tasks previously performed by humans — people worry about their place in the organization, their relevance to the team, and whether their career trajectory has just been disrupted. These are not hypothetical fears generated by people who have not thought carefully about their situation. They are rational responses to a landscape in which automation is genuinely changing the value proposition of certain skills.

These fears deserve honest engagement. The manager who dismisses them with cheerful assurances ("AI is just a tool, it won't replace anyone!") will lose credibility immediately with any team member who has read the news in the past two years. The manager who validates them without context ("yeah, this industry is changing fast, who knows what's coming") creates the very anxiety they need to reduce. Neither response helps the team member understand their actual situation or identify a productive path forward.

The productive middle ground is honest, mission-aligned framing. AI tools change the nature of work; they do not eliminate the need for judgment, relationships, accountability, and leadership. The team member who spent four hours per week generating manual reports now has four hours for higher-value analytical work. The team member who spent time on repetitive communications can now focus on complex stakeholder conversations. Reframing AI adoption in concrete, team-specific terms transforms the conversation from an existential threat to an operational opportunity. This reframing is not spin — it reflects the actual trajectory of well-managed AI adoption in high-performing teams.

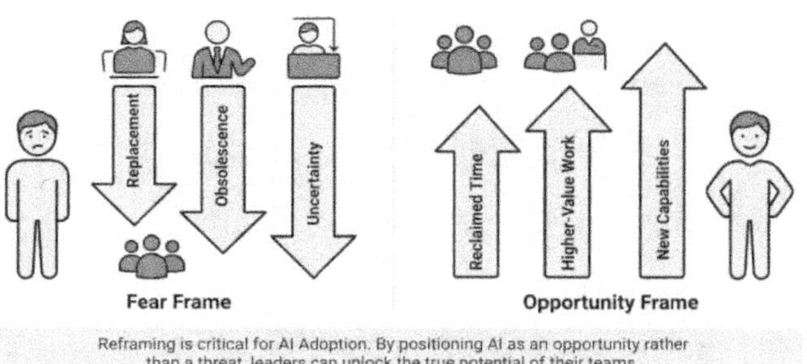

Reframing is critical for AI Adoption. By positioning AI as an opportunity rather than a threat, leaders can unlock the true potential of their teams.

14.5.1 The Fear Inventory

One of the most effective early moves in leading AI adoption is conducting an explicit, structured inventory of your team's concerns. This is not a suggestion box or an anonymous survey — it is a direct conversation, ideally in a one-on-one format, where you ask each team member what specifically worries them about the change. The direct conversation format matters: anonymous surveys produce surface-level themes; direct conversations produce the specific, personal concerns that are actually driving behavior.

The fear inventory serves multiple purposes. It provides accurate intelligence on where resistance is concentrated, what training gaps exist, and which individual concerns require targeted attention. It demonstrates to each team member that their concerns are heard rather than dismissed.

And it shifts the conversation from passive anxiety to active articulation — once a fear is named, it becomes something that can be addressed rather than something that festers. The act of naming a fear also reduces its psychological power: the concerns that seem overwhelming in the abstract become more manageable when stated specifically.

After conducting individual fear inventories, synthesize the themes without attributing specific concerns to individuals, and bring the synthesized concerns back to the team in a facilitated discussion. This process normalizes the experience of concern, demonstrates that others share similar anxieties, and creates shared ownership of the problem-solving process. The follow-up discussion should produce concrete actions — not reassurances — that address the specific concerns identified. When team members see their concerns translated into actual management action, trust in the adoption process increases substantially.

14.5.2 Building Enthusiasm Through Early Wins

Fear recedes fastest when it is replaced by capability. The most powerful antidote to AI anxiety is the experience of using a tool effectively and finding that it genuinely makes something easier. As a manager, your job is to engineer that experience as early as possible for every member of your team. The sequence matters: early wins before broad deployment, not as a reward for adoption but as a prerequisite for it.

Identify the lowest-friction, highest-impact AI use case for your team's specific work. This is not necessarily the most sophisticated capability — it is the one that provides an immediate, tangible benefit with minimal learning curve. Schedule a structured hands-on session for the team to work through this use case together, with support available. Celebrate the first person who uses the tool to produce something genuinely useful. Make the early adoption visible and positive, not competitive or evaluative. An atmosphere of shared discovery is more conducive to adoption than a performance assessment environment.

The pilot-to-production pathway is the structured approach to this progression: start with a contained pilot involving willing early adopters, measure results rigorously, share them transparently, and use the evidence to build confidence among more hesitant team members. This sequence converts theoretical enthusiasm into organizational momentum. The keyword in "pilot-to-production" is the arrow: pilots that are not designed to inform production decisions are demonstrations, not investments. Every pilot should have a defined set of conditions under which it will be scaled up, and the evaluation criteria should be set before the pilot begins.

14.6   Managing the Human Side of Digital Transformation

Digital transformation is a business strategy executed by human beings. The technical architecture of a transformation — the platforms selected, the integrations built, the data pipelines configured — is the part that is planned, budgeted, and tracked on implementation dashboards. The human architecture — the trust, engagement, capability, and morale of the people implementing and using the technology — is most often underestimated and most frequently responsible for failure. This asymmetry between investment in technical planning and investment in human change management is one of the most consistent findings in transformation research.

As a team manager, you do not control the technical architecture. But you have significant influence over the human architecture, and that influence is proportional to your intentionality. Managers who approach digital transformation with the same rigor they apply to project management — regular check-ins, documented concerns, specific interventions, tracked outcomes — consistently produce better team-level results than managers who rely on organizational change programs to handle the human dimension. Organizational change programs create the enabling conditions; team-level management creates the actual outcomes.

Diagram 13.6: The Human Architecture of Digital Transformation

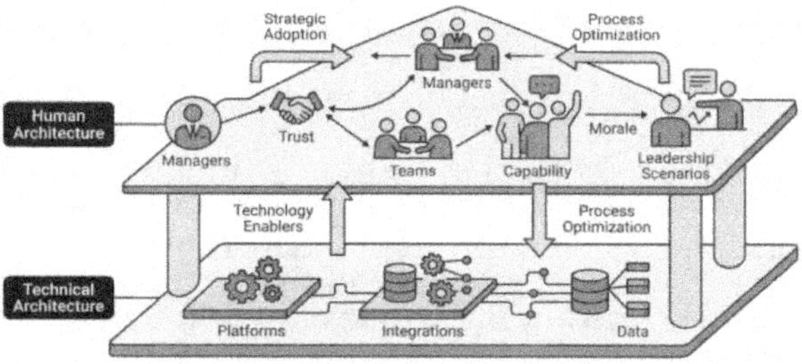

A robust technical foundation only realizes its full value when activated by a motivated and capable human architecture.

14.6.1 Identifying and Supporting At-Risk Team Members

Not all team members experience change equally. Some people are natural adapters — they embrace disruption, learn quickly, and often become informal advocates for new tools. Others are constitutionally more risk-averse, more attached to established ways of working, or more personally invested in the skills that the new tools are changing. Neither type is better or worse as an employee; both types need different things from you as their manager during a transformation.

Understanding this spectrum within your specific team and acting on that understanding is the practice.

Your at-risk team members — those most likely to disengage, underperform, or leave during a transformation — are not always the ones who complain loudest. Sometimes they are the quiet ones, the long-tenured employees whose identity is tightly tied to a way of working that is now changing, or the high performers who feel threatened by a shift that reduces the relative advantage of their existing skills. Learning to identify at-risk team members early — through one-on-one conversations, behavioral shifts, output changes, and direct observation — allows you to intervene before disengagement becomes departure. The observation skills required here are the same ones that make great coaches: noticing what is different, not just what is wrong.

Your intervention toolkit includes direct conversations, targeted development support, role evolution conversations, and, in some cases, honest discussions about fit. Protecting a team member from change they genuinely cannot or will not make is not kindness — it delays a conversation that becomes more painful the longer it waits. But providing the resources, time, and psychological safety to enable genuine adaptation is a core responsibility of management. Most people who appear resistant to change are actually resistant to the uncertainty, skill gap, or perceived threat that change represents — and when those underlying factors are addressed, resistance typically resolves.

14.6.2 Preserving Institutional Knowledge During Transitions

One of the most consequential and least discussed risks in digital transformation is the loss of institutional knowledge. When tools change, some of the expertise required to use the old tools becomes obsolete. But embedded within that expertise is often a deeper knowledge of business processes, customer relationships, domain history, and organizational context — that is not embedded in any tool and cannot be replaced by any new platform. This deeper knowledge is the true institutional memory of the organization, and it is most concentrated among team members who have been with the organization the longest.

When experienced team members disengage or leave during a transformation, that deeper knowledge often leaves with them. As a manager, your job is to identify the institutional knowledge most at risk and create mechanisms for its preservation before the transition is complete. This might mean structured knowledge-transfer sessions, documentation sprints, mentoring pairings between long-tenured and newer team members, or simply deliberate conversations that make tacit knowledge explicit. The urgency of this work is inversely proportional to how explicitly it is programmed into the transformation plan — that is, it rarely receives the attention it deserves unless someone makes it a priority.

AI tools can actually accelerate this preservation effort. Meeting transcription and summary tools capture institutional knowledge that was previously exchanged only verbally. AI-powered knowledge base systems can organize and make searchable the documentation your team produces. The technology that is disrupting your team's existing ways of working can also help you preserve what should not be lost in the transition. This is one of the clearest examples of AI serving the organization's human goals rather than simply replacing human functions.

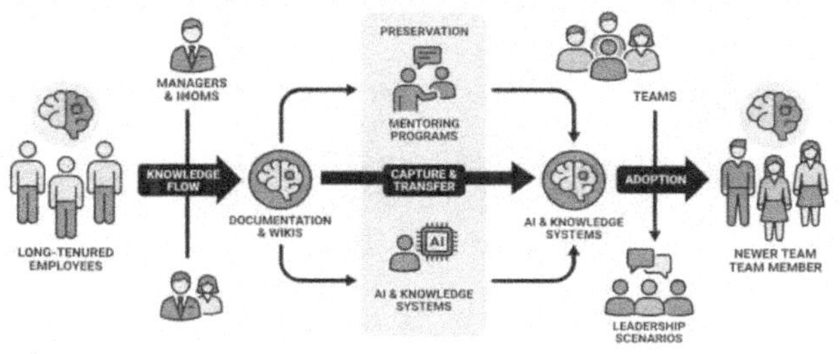

A structured approach to capturing and transferring critical knowledge ensures organizational resilience and continuity during periods of transition.

14.7 Supporting Your Team Through Uncertainty and Ambiguity

The hardest conversations during transformation are not the ones where you have bad news to deliver — those at least have a clear subject. The hardest conversations are the ones where the honest answer is "I don't know." Will the platform be stable by the deadline? I don't know. Will team size change after the transition? I don't know. Will the new roles look significantly different? I don't know. The impulse to provide an answer — any answer — in these moments is strong, because the social pressure to project confidence is real and because admitting uncertainty feels vulnerable in a way that is uncomfortable for many managers.

Uncertainty is not a management failure. But handling uncertainty poorly is. The manager who pretends to have answers they do not have loses credibility the moment reality

376

diverges from their prediction. The manager who retreats into silence leaves a vacuum that rumor fills. The manager who acknowledges uncertainty while providing structure, continuity, and genuine presence gives the team what they actually need: a stable anchor during an unstable period. The research on uncertainty management in organizational settings consistently shows that what people most need during periods of high ambiguity is not answers but presence — the experience of a leader who is visible, engaged, and honest.

Three practices anchor effective uncertainty management. First, separate what you know from what you don't know, and communicate both categories explicitly. "Here is what I can confirm: the timeline is Q3 and the platform is locked. Here is what I don't know yet: how our specific workflows will be affected. I'm working to find out and will update you by next week." Second, maintain operational consistency in everything you control. If one-on-ones stay on the calendar, if deadlines are still respected, if your presence and availability remain consistent, you communicate that not everything is in flux — just the specific things that are. Third, focus the team's attention on the near term. When the long-term is genuinely uncertain, directing energy toward the next two weeks — the concrete deliverables, the immediate milestones, the specific skills to develop right now — keeps the team productive and prevents the paralysis that comes from staring at a fog-covered horizon.

14.8 Building Adaptive Teams That Thrive in Continuous Change

The goal of change management is not to get your team through this particular transformation intact. It is to build the adaptive capacity that allows your team to move through the next transformation more fluidly — and the one after that, and the one after that. The world your team will operate in for the next decade is one of continuous change, and a team that survives each change only by expending maximum effort will eventually run out of capacity. The investment in adaptive capacity is therefore not a performance management initiative — it is a sustainability strategy.

Adaptive capacity is not a personality trait that some people have and others lack. It is a skill set that can be developed, and a team condition that can be cultivated. Managers who build adaptive teams do several things consistently: they celebrate learning at least as much as they celebrate output, which signals that the process of acquiring new capability is organizationally valued; they give team members regular exposure to new challenges rather than protecting high performers with comfortable specializations; they debrief transitions explicitly, asking "what did we learn about how we handle change?" rather than simply moving on; and they create psychological safety around the admission of confusion, which allows people to ask for help before they fall behind. Each of these practices contributes to a team culture that frames change as navigable rather than threatening.

Diagram 13.8: Building Adaptive Team Capacity

Each change cycle strengthens the team's ability to handle the next one.

The manager's role in building this adaptive capacity is partly what you do and partly what you model. If you approach the transformation you are leading with curiosity rather than defensiveness, if you acknowledge your own learning process openly, and if you demonstrate that uncertainty is navigable rather than catastrophic, you give your team implicit permission to adopt the same stance.

Leadership in times of change is fundamentally contagious — and that works in both directions. A genuinely curious manager, genuinely honest about what they do not know, and genuinely committed to the team's success in navigating the transition creates conditions in which the team can do the same. That is the foundation of operational clarity amid ambiguity.

14.9 Manager's Checklist for Leading Change

Use this checklist at the outset of any significant organizational or team-level change initiative and revisit it monthly throughout the transition.

- Diagnose where each team member sits on the ADKAR scale before deploying broad change programming. Customize your support for each individual based on whether the primary gap is awareness, desire, knowledge, ability, or reinforcement.
- Schedule explicit change communication into your existing meeting cadence rather than relying on organizational announcements to carry the message. Own the local communication channel.
- Conduct a one-on-one fear inventory with every team member within the first two weeks of a major

change announcement. Document the themes and address them in a team-level facilitated discussion.

- Identify your early adopters and engineer visible, low-stakes early wins that can build confidence across the broader team through the pilot-to-production pathway.
- Identify your at-risk team members — not just the loudest objectors, but the quiet ones whose behavioral signals suggest disengagement — and develop individualized support plans.
- Initiate explicit institutional knowledge preservation efforts before long-tenured team members have a chance to disengage. Pair knowledge transfer with AI-assisted documentation tools to capture what exists only in people's heads.
- Separate what you know from what you do not know in every transformation communication, and commit to specific timelines for resolving open questions.
- Maintain operational consistency in everything within your control — one-on-ones, deadlines, norms, your own presence — to signal that not everything is uncertain.
- Debrief every significant transition formally, asking not just "did we deliver?" but "what did we learn about how we handle change?" and incorporate answers into the next transition plan.

14.10 **The Path Forward**

Leading through change is among the most demanding and most consequential work a manager does. It requires technical literacy, human sensitivity, communicative courage, and operational discipline — all at once. It also requires the ability to operate in ambiguity without projecting that ambiguity onto your team as anxiety. This combination of demands is exactly what makes leading through transformation one of the clearest signals of real leadership capability — it is not possible to fake your way through it with charisma or expertise alone.

The frameworks in this chapter — Kotter, ADKAR, the emotional arc, the manager as translator role, the fear inventory, the pilot-to-production pathway, the adaptive team model — are not scripts. They are lenses through which to interpret what you are seeing and tools for responding. Effective change leadership is not the application of a single model; it is the capacity to move fluidly between frameworks as the situation evolves. A team member in the Denial stage of the change curve needs a different response than one stuck in the Knowledge gap of ADKAR. A Kotter-style urgency conversation is more appropriate at the start of a transformation than in month four, when the focus shifts to sustaining momentum rather than generating it.

What remains constant, regardless of which framework you apply, is this: your team will go where you go. Not because they agree with you, not because they are required to, but because in the absence of a visible, confident, honest leader, uncertainty fills the space and people retreat to self-protection. Be the person who walks forward with honest acknowledgment of what is unknown and a genuine belief in the team's ability to navigate it. That posture, more than any framework or checklist, is what leading through change actually looks like — and it is the posture that builds the trust, the adaptive capacity, and the organizational resilience that make future transformations easier to lead.

15 Building Team Culture and Driving Engagement

15.1 Scenario

You inherited the team six months ago. On paper, it looks functional: deliverables are mostly on time, no one has filed a complaint, and turnover has been flat. But something is missing. Meetings are transactional. People do their work and disconnect. Nobody pushes back on bad ideas, but nobody contributes unexpected good ones either. The team produces exactly what is asked of it and nothing more — no initiative, no energy, no visible pride in what gets made. You have delivered results, but you know the team is operating at a fraction of its actual capacity.

You bring this observation to a peer manager over coffee, and she offers the bluntest possible diagnosis: "Your team doesn't have a culture. They have a routine." She is right, and the distinction matters more than most management textbooks let on. A routine is a pattern of behavior that persists because nothing has disrupted it. A culture is a pattern of behavior that persists because people believe in something together — because they share a sense of who they are, what they value, how they treat each other, and why the work matters. You cannot mandate culture into existence. But you can create the conditions in which culture grows, and you can sustain them intentionally over time.

This chapter is about doing exactly that — building a team culture that is genuine rather than performative, durable rather than dependent on a single initiative, and powerful enough to drive real engagement even when the work is hard. It is written with the understanding that you did not inherit a blank slate: you inherited a team with existing norms, histories, and dynamics. Culture-building in that context is not starting from zero — it is deliberate, patient cultivation of the conditions that shift a team from functional compliance to genuine engagement.

15.2   Why It Matters

Culture is not a soft variable. It is a performance variable. The research is unambiguous: teams with strong psychological safety produce more creative solutions, surface problems earlier, recover from failures faster, and retain talent at significantly higher rates than teams without it. Engagement, which is culture's most direct measurable output, correlates powerfully with productivity, quality, and customer outcomes. A disengaged team member costs the organization an estimated 34 percent of their annual salary in lost productivity — every year, compounding, until they either reengage or leave. Multiply that across a team of eight to twelve people, and the business case for culture investment becomes immediate.

For new managers, team culture also serves as a credibility signal. The state of your team's culture reflects directly on your management — not because you are responsible for every interpersonal dynamic, but because culture is the accumulated residue of the decisions, norms, and behaviors you model, permit, and reinforce over time. A team with a genuinely strong culture gives its manager a platform to do ambitious work. A team without one spends its collective energy managing internal friction rather than producing outcomes. The difference between these two states is not accidental — it is the sum of hundreds of small management decisions made or unmade over months and years.

The AI dimension of culture-building is newer but equally consequential. As AI tools enter team workflows, they alter collaboration patterns, change the nature of certain roles, and raise questions about equity in the distribution of automation benefits and burdens. Teams with strong, inclusive cultures absorb these changes without fracturing. Teams with fragile cultures experience them as additional stressors that amplify existing tensions. Building culture deliberately — before the next disruption arrives — is an act of organizational risk management, not just people management. It is the team-level equivalent of building technical resilience before the system is under load.

Diagram 14.2: Culture as a Performance Variable

When teams feel psychologically safe, engagement soars, driving exceptional performance outcomes.

15.3 Defining and Shaping Your Team's Culture

Culture is not discovered — it is designed and maintained. It does not have to be designed by a consultant or articulated in a formal workshop, but it must be intentional. Teams that have great cultures almost always have a manager who made deliberate choices about what to model, what to celebrate, what to tolerate, and what to refuse. The absence of deliberate cultural design does not produce a neutral environment — it produces a culture shaped by defaults: the most assertive voices, inherited norms from previous managers, and the path of least resistance. Deliberate design is not optional if you want a culture that is genuinely yours.

15.3.1 The Four Cultural Levers

Four levers shape team culture more powerfully than any others: what the manager models, what the manager celebrates, what the manager tolerates, and what the manager refuses. These four levers operate continuously, whether you are aware of them or not, and your team reads them with extraordinary precision. They are watching not just what you say but what you do, what you allow, what you recognize, and what you refuse to permit — and they calibrate their own behavior accordingly, often without being consciously aware of it.

What you model is the most powerful lever of all. If you are chronically late to meetings, your team will interpret punctuality as optional. If you respond to every question with a defensive explanation, your team will learn that admitting uncertainty is risky. If you speak dismissively about adjacent teams, your team will do the same. Conversely, if you acknowledge mistakes openly, express genuine curiosity, treat every team member with consistent respect regardless of their role or seniority, and take care of your own well-being visibly enough for your team to see it is permitted, you communicate the behaviors that are truly valued. The clarity of modeling as a signal is precisely why leadership is described so often as a visibility job — your behaviors are always on display, whether you intend them to be or not.

What you celebrate is the second lever. Recognition sends a cultural signal: this behavior, this outcome, this way of working is what we want more of. Managers who only recognize deliverable completion send a message that results are all that matter — and inadvertently discourage the risk-taking, collaboration, and candor that produce great results over time. Managers who also recognize the team member who flagged a problem before it became a crisis, the person who helped a colleague without being asked, and the individual who suggested a better approach and dared to push for it — those managers are actively building the culture they want. Recognition is not administrative — it is cultural architecture.

What you tolerate is the third lever — and the one most likely to undermine a manager's cultural intentions. Every team norm you allow to violate your stated values without consequence sends a message that those values are aspirational, not operational. A manager who says the team values psychological safety but never addresses a colleague who routinely mocks others' ideas in meetings communicates that psychological safety is something the team talks about, not something it actually practices. Culture is not what you say you value. It is what you consistently hold the line on. The line is visible because of what happens when it is crossed.

What you refuse is the fourth lever. The willingness to say no — to a request that would harm your team, to a behavior that conflicts with the team's values, to a norm that has crept in without being consciously chosen — is a cultural act. Every time you refuse to accept something that conflicts with who your team is and what your team stands for, you are making the culture more real. Refusals are credibility moments: they demonstrate that the stated values have behavioral consequences, not just rhetorical ones.

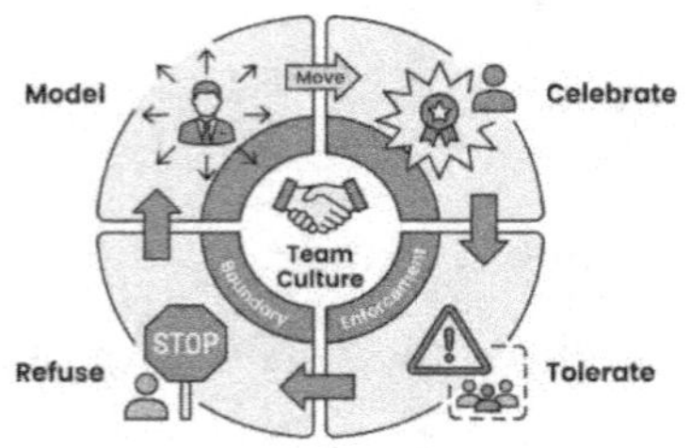

Culture is actively shaped by these four management behaviors.

15.3.2 Articulating Your Team's Values

Culture becomes more durable when it is made explicit. Teams that have names for their values, shared language for their norms, and deliberate rituals that embody their identity are more resilient than teams that operate on an unspoken cultural consensus that only the longest tenured members fully understand. When new team members join, explicit values can be communicated directly. When conflict arises, explicit values provide a reference point. When decisions need to be made under pressure, explicit values create operational clarity that saves time and reduces friction.

Facilitating a values articulation exercise does not require an off-site retreat or an organizational development consultant. In a regular team meeting, ask the team three questions: What kind of team do we want to be? How do we want to treat each other when the work gets hard? What would an outsider say about this team that we would be proud of? The answers to these questions, synthesized and stated back in plain language, become your working culture statement. The synthesis is as important as the questions — the manager's role is to reflect the team's answers in language that is specific, honest, and genuinely theirs rather than borrowed from a corporate values brochure.

The critical step after articulation is activation. A list of values on a slide deck is decorative. Values referenced in hiring conversations, onboarding discussions, project retrospectives, conflict resolution, and recognition — values used as operating language, not display language — are alive. Every time you say "that's exactly the kind of initiative we celebrate here" or "that's not consistent with how we've agreed to work together," you are activating the culture rather than simply displaying it. Activation is the practice that converts intention into behavior, and behavior into culture.

15.4 Inclusion, Belonging, and Psychological Safety

The most well-researched predictor of high-performing teams is not average talent level, compensation structure, or tool sophistication. It is psychological safety — the shared belief that the team is safe for interpersonal risk-taking. Teams with high psychological safety speak up when they see problems, offer ideas without fear of ridicule, admit mistakes quickly, and disagree productively. Teams without it optimize for self-preservation, and organizational learning grinds to a halt. This finding — robust across industries, organizational sizes, and cultural contexts — is arguably the most practically important piece of team science available to managers.

15.4.1 Building Psychological Safety

Psychological safety is not a mood. It is a structural property of a team environment, built through repeated behavioral evidence that taking interpersonal risks is safe. Building it requires consistency over time and deliberate action in the moments that matter most. It also requires patience: teams that have operated in psychologically unsafe environments for years do not become safe after a single team-building exercise or a new manager's declaration that things will be different. They become safe through months of accumulated evidence.

The moments that matter most are the moments when someone does something brave: offers a dissenting opinion, admits they made an error, asks what might seem like a basic question, or raises a concern about a decision that has already been made. How you respond in those moments — not in your culture workshop, not in your team values document, but in the actual moment — determines whether the team learns that psychological safety is real or aspirational. Your response in the brave moment is the evidence the team uses to calibrate its actual safety level.

When someone admits a mistake, respond with curiosity rather than judgment: "What happened? What did we learn? What would we do differently?" When someone raises a dissenting opinion, treat it as data rather than a problem to manage: "That's worth exploring. Tell me more about what you're seeing." When someone asks a question that reveals they are behind, create space rather than closing it: "Good that you asked — let me make sure everyone has what they need to answer that question." These responses are not scripts — they are orientations toward the team member's brave act that signal the act was worth taking.

Inclusion is the systemic dimension of psychological safety. A psychologically safe team is not one where everyone always feels comfortable — it is one where the discomfort of risk-taking is distributed equitably. If only some team members feel safe enough to dissent, challenge, or admit uncertainty, the team is not actually psychologically safe — it is simply comfortable for the majority while excluding others from full participation. The manager's job is to ensure that safety is not reserved for those with the most social capital, the most seniority, or the most similarity to the manager's own background and style.

Inclusion in practice means paying attention to whose voices are centered and whose are peripheral in team discussions, making explicit space for input from quieter team members, examining whether your recognition patterns consistently favor certain profiles or styles, and creating structural mechanisms — rotating facilitation, anonymous input channels, explicit time for dissenting perspectives — that compensate for the natural dynamics that privilege certain voices over others. These structural mechanisms are not substitutes for inclusive behavior — they are supports that make inclusive behavior more consistent and equitable.

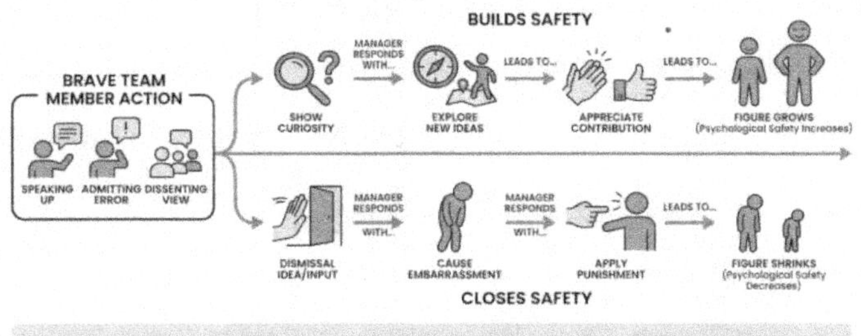

15.5 Employee Engagement Drivers

Engagement is the behavioral expression of culture. An engaged team member is not simply a happy one — engagement is the condition in which people bring discretionary energy and genuine commitment to their work, going beyond the minimum required because they genuinely want to contribute to the team's success. Understanding what drives engagement at the individual level — what makes your specific team members bring that extra energy — is among the most practical management skills you can develop. It requires ongoing attention because the drivers shift as people's work, relationships, and life circumstances evolve.

The research on engagement consistently identifies three core drivers: autonomy, mastery, and purpose — a framework most closely associated with Daniel Pink's work on intrinsic motivation. Each driver operates at the individual level, which means your role is to understand how each team member experiences each driver and tailor your management accordingly. The team member who is deeply driven by mastery needs a different conversation than the one whose primary driver is purpose. Generic engagement programs that treat all team members as having the same motivational profile routinely underperform targeted individual engagement conversations.

15.5.1 Autonomy

Autonomy is the experience of having meaningful control over one's work. It is not unlimited freedom — it is the right balance of direction and discretion that allows people to bring their judgment and creativity to their work. Managers who over-specify how work should be done — providing detailed instructions for tasks that team members are fully capable of approaching independently — send a message that judgment is not trusted, and engagement declines accordingly. The pattern is consistent: the more a manager specifies methods rather than outcomes, the lower the team's initiative and creative contributions.

Building autonomy into your team's experience means defining outcomes clearly and leaving the approach flexible, where the risk of variation is acceptable. It means resisting the impulse to review work at every stage and instead agreeing on check-in points that allow for meaningful progress between your engagements. It means creating space for team members to propose how they would tackle a problem rather than always presenting a method for them to execute. The quality of the output often improves when the person doing the work has shaped the approach — not because the manager's approach was wrong, but because personal ownership of method generates additional investment in quality.

In AI-augmented teams, autonomy takes on an additional dimension. Team members who are given the freedom to experiment with AI tools — to discover which capabilities are genuinely useful in their specific work — build ownership of the technology rather than experiencing it as something imposed on them. Forced adoption without exploration of autonomy is one of the most common failure modes in AI tool rollouts. The team member who discovered a use case for themselves will deploy it with significantly more skill and consistency than one who was trained on a use case their manager identified.

15.5.2 **Mastery**

Mastery is the experience of getting better at work that matters. People who feel they are growing — who experience the satisfaction of developing a new skill, solving a harder problem, contributing at a higher level — are engaged. People who feel their work has plateaued, or who are kept in roles that do not challenge them, disengage. The disengagement from stagnation is often quiet — it does not present as complaint or conflict but as diminished energy, reduced initiative, and eventually the kind of functional compliance that produces your outputs but nothing more.

The manager's role in mastery is primarily one of challenge calibration. Your job is to understand each team member's current capability level, identify the specific growth edges that would most benefit them and the team, and provide assignments that stretch without overwhelming. The goal is the productive difficulty described in learning science as the "zone of proximal development" — hard enough to require real effort and real learning, achievable enough that success remains possible. This calibration requires genuine knowledge of each team member's current capabilities and growth trajectory, which is why regular, substantive one-on-one conversations are not optional for managers who want to build mastery-driven engagement.

AI tools create unprecedented opportunities for accelerated mastery. AI-powered learning platforms can deliver personalized content matched to each team member's current skill level and learning style. AI simulation tools allow people to practice complex skills in low-stakes environments before deploying them in real situations. And the efficiency gains from AI automation in routine tasks create time that can be redirected toward the deeper, more complex work that builds genuine expertise. The net effect of well-implemented AI adoption on mastery-driven engagement can be substantially positive — but only if the reclaimed time is actually directed toward growth rather than absorbed by additional volume of the same work.

15.5.3 Purpose

Purpose is the experience of contributing to something that matters beyond the immediate task. Purpose does not require heroic language or organizational mission statements — it requires genuine line-of-sight between daily work and meaningful impact. A customer service team member who understands precisely how their response quality affects customer retention and how customer retention affects the business finds purpose in answering support tickets. A developer who understands how their code change affects real users finds purpose in debugging sessions. The work is the same; the experience of purpose depends on whether the connection to impact is visible.

Your role in building purpose is to make the connection visible and specific. Do not rely on the organizational mission statement to provide it — it is too abstract for daily motivation. Create a direct connection between what your team does and who it affects: bring in a customer to share their experience, share specific feedback you received about the team's work, celebrate outcomes, not just outputs, and help each team member see the downstream consequences of their contribution. When team members understand who depends on their work and how their work affects those people, the motivational quality of otherwise mundane tasks changes substantially.

Diagram 14.5: The Three Engagement Drivers Applied to AI-Augmented Teams

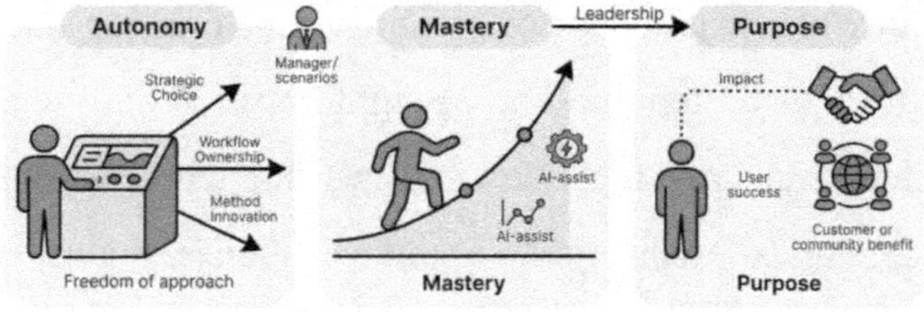

Leveraging Autonomy, Mastery, and Purpose empowers teams to thrive in an AI-enhanced work environment.

15.6 Managing Hybrid and Remote Teams Effectively

Hybrid and remote work have permanently reshaped the team environment. The informal culture infrastructure that developed organically in physical offices — the hallway conversation, the shared lunch, the visible body language cue that told you something was wrong before anyone said a word — does not translate automatically to distributed teams. Managers who treat hybrid and remote teams as simply "office teams but on video" consistently underperform those who redesign their cultural practices for the medium. The redesign is not about replicating what worked in person; it is about building intentional structures that produce the same outcomes — connection, trust, shared identity — through different means.

15.6.1 Designing for Distributed Connection

Distributed connection requires deliberate design. It does not happen because the right people are on the right Slack channels. It happens because the manager creates structures that produce the types of interaction — substantive, human, trust-building — that build team cohesion over time. Designed interactions in distributed space must replace the spontaneous interactions that occur naturally in shared physical space, and the design must be intentional enough to produce the same quality of human connection that spontaneous interaction produced organically.

The most effective distributed connection practices share a common characteristic: they provide structure without scripting. A standing "wins and learnings" segment at the start of a team meeting gives people a reason to share something human and genuine without requiring it to be spontaneous. A rotating "spotlight" practice — where each team member takes two minutes to share something about their work, a challenge they solved, or something they are working on — builds mutual visibility without mandating vulnerability. An asynchronous appreciation channel where people can share recognition publicly creates a record of the team's culture that is visible to everyone, including new team members who joined after the events being recognized.

The manager's physical presence in hybrid teams requires specific attention. If you are in the office on some days and on video on others, your behavior on both must signal the same cultural messages. Informal conversations in the office must not become the exclusive channels for decision-making that exclude remote team members. Virtual team members must receive equivalent access to development opportunities, project assignments, and visibility into organizational dynamics. Equity in a hybrid context is not about identical treatment — it is about equivalent opportunity, and maintaining it requires active attention rather than passive goodwill.

Diagram 14.6: Hybrid Team Connection Design

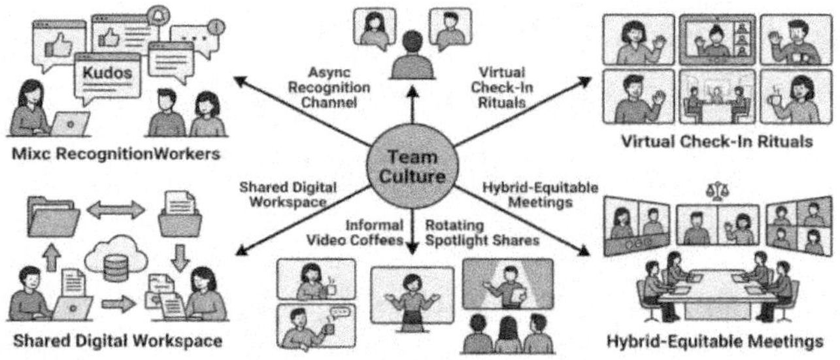

Cultivating a cohesive team environment through deliberate distributed connection practices.

15.6.2 Equity Across Hybrid Configurations

The research on hybrid work reveals a consistent proximity bias: managers rate in-office employees higher on performance evaluations, assign them to more visible projects, and include them more frequently in informal decision-making — even when remote employees are performing equivalently. This bias is not usually deliberate, but its effects on culture are serious: remote employees who experience it disengage and eventually leave, depriving the team of talent that was performing well. Proximity bias is particularly damaging because it is self-reinforcing: as in-office employees receive more opportunities, their performance improves relative to remote colleagues, who receive fewer, creating an apparent performance difference that validates the original bias.

Counteracting proximity bias requires structural interventions, not just good intentions. Rotate the location of team meetings so that different configurations — fully remote, hybrid, fully in-person — occur regularly. Explicitly document decisions that emerge from hallway conversations and share them with the full team. Track your project assignments and development opportunities to verify that they are not systematically favoring proximity. And in performance evaluations, anchor assessments in objective outcomes rather than subjective impressions of engagement and presence. These structural interventions do not resolve every dimension of hybrid equity, but they remove the most common structural mechanisms through which bias operates.

15.7 Preventing Burnout and Supporting Well-Being

Team culture and team well-being are inseparable. A culture that does not protect sustainable work is not healthy, regardless of how strong the other indicators look. Burnout is not a personal failure — it is a systemic condition that emerges when people are consistently required to spend more than their sustainable energy reserves without adequate recovery. A team that produces excellent results for two years while its members gradually deplete is not a high-performing team — it is a team in the process of consuming itself. The eventual failure will be sudden and costly.

As a manager, your role in burnout prevention is both protective and modeling. Protective actions include monitoring workload distribution to ensure no team member is consistently absorbing unsustainable volume, creating explicit norms around off-hours availability and communication, using workload data and AI project management tools to make invisible overload visible, and treating rest as a legitimate productivity strategy rather than a sign of reduced commitment. These protective actions create the structural conditions for sustainable work; they do not substitute for the cultural permission actually to use them.

Modeling is equally important. A manager who sends emails at midnight and never takes a vacation is not demonstrating dedication — they are communicating that those behaviors are the standard. If you want your team to have sustainable work habits, you must visibly practice them yourself. Block focus time on your calendar and protect it. Use your vacation time fully. Speak openly about protecting your energy as a leadership responsibility, not a personal indulgence. The cultural permission to work sustainably flows from the top of the team hierarchy; if that permission is absent from your behavior, it will be absent from the team's behavior, regardless of what you say about work-life balance in team meetings.

15.7.1 Using AI Pulse Surveys and Sentiment Analysis

One of the most valuable AI applications for culture and engagement management is the modern pulse survey combined with sentiment analysis. Traditional employee engagement surveys were run annually, produced results months after data collection, and were too coarse to surface team-level dynamics. AI-powered pulse tools run continuously, aggregate responses in real time, and surface patterns that would be invisible in manual analysis. They are not a substitute for direct management presence, but they are a powerful supplement — particularly for surfacing concerns that team members are reluctant to raise in direct conversation.

As a manager, these tools provide a data layer that supplements your direct observations. They surface concerns that team members may not raise in one-on-ones, identify trends before they become crises, and provide evidence that you can bring to leadership conversations when requesting resources or raising concerns. They also provide an accountability mechanism: when the data shows that team sentiment has declined on a specific dimension, you have both the visibility to see it and the obligation to address it. The transparency of pulse data — the fact that trends are visible in near-real time — creates a form of organizational accountability that annual surveys cannot produce.

Use pulse data as intelligence, not as a scorecard. Low sentiment scores are not a judgment on you — they are information about what needs attention. The manager who responds to declining pulse scores with defensiveness misses the point and loses the tool's trust. The manager who brings the data into an honest team conversation — "I'm seeing some signals in our recent pulse that suggest people are feeling stretched. Let me share what I'm seeing and hear what's behind it" — uses the data as it was intended: as a catalyst for genuine dialogue and targeted action rather than as a performance indicator to be managed.

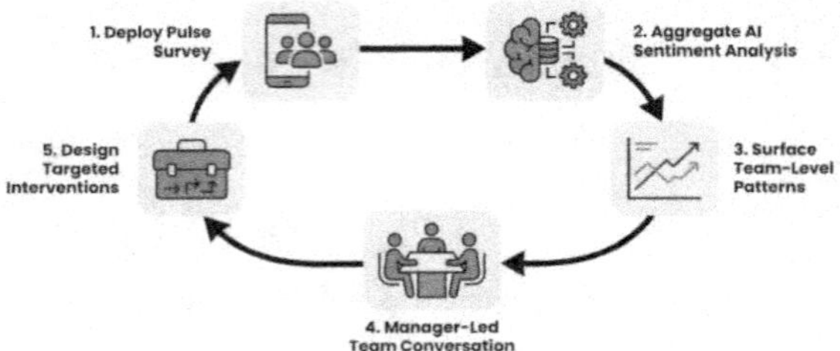

15.8 Celebrating Wins and Building Rituals That Stick

The rituals a team practices regularly are the most visible expression of its culture. Rituals are not team-building exercises — they are the repeated behaviors that encode what the team believes, how the team relates, and what the team values. A team that gathers regularly to celebrate wins believes that wins are worth celebrating. A team that ends every project with a structured retrospective believes that learning is worth examining. A team that starts every Monday with shared context believes that alignment is worth the time investment. The specific rituals matter less than their consistency and authenticity: a ritual performed perfunctorily becomes noise, but one practiced with genuine engagement becomes part of the team's identity.

15.8.1 Designing Recognition That Resonates

Recognition is among the most powerful cultural tools available to a manager, and among the most frequently misused. Generic recognition ("great job everyone, really appreciate the hard work") is better than silence, but not by much. Specific recognition — tied to a concrete behavior, outcome, or decision that aligns with the team's values — is qualitatively different. It tells the recognized person that you saw what they did. It tells the team what the valued behaviors look like in practice. And it reinforces the culture by making abstract values concrete. The specificity is not a stylistic preference — it is the mechanism through which recognition actually works.

The most effective recognition systems combine public and private elements. Public recognition — in team meetings, in shared channels, in communications that reach leadership — provides visibility and signals organizational value. Private recognition — a direct message or one-on-one conversation that acknowledges the specific behavior without an audience — communicates personal attention and genuine appreciation. Both are necessary; neither is sufficient alone. Public recognition without private acknowledgment can feel performative; private recognition without a public dimension misses the opportunity for cultural transmission.

Recognition must also be equitable. Research consistently shows that recognition patterns in most teams systematically overrepresent extroverted team members, those whose contributions are most visible, and those who are most similar to the manager in working style and background. Building equity into recognition means actively tracking who has been recognized recently, making deliberate efforts to surface the contributions of team members who do valuable work quietly, and creating recognition mechanisms — peer appreciation, structured spotlights, rotating contribution highlights — that are not dependent on the manager's selective attention. The manager's attention is inevitably biased; structural mechanisms compensate for that bias.

15.8.2 Building Rituals That Stick

The rituals that endure in team culture share three characteristics: they are simple enough to sustain, they encode something genuine, and they are owned by the team rather than imposed by the manager. Complexity is the enemy of ritual consistency: a recognition practice that requires fifteen minutes of preparation each week will be skipped more often than one that requires two minutes. Authenticity is the enemy of ritual emptiness: a ritual that does not reflect something the team actually believes will feel hollow within months and will be abandoned or performed without meaning. Ownership is the enemy of ritual fragility: a ritual that belongs to the manager leaves with the manager.

Simple rituals that encode genuine values and that the team has helped design are remarkably durable. They persist through personnel changes, through leadership transitions, through organizational disruptions — because they are not tied to any individual. They are part of how the team defines itself. When a new team member joins, they encounter these rituals as part of the team's identity and adopt them as their own. When the manager transitions out, the rituals continue because they belong to the team, not to the manager.

Design your team's rituals collaboratively. Ask the team: "What regular practices would make us feel more connected, more recognized, or more aligned?" The answers will surprise you with their specificity and practicality. A team that identifies "we want to spend five minutes at the start of every Monday call sharing one thing that went well last week" is more likely to sustain that practice than a team that was instructed to do it by a manager following a playbook. The participation in design is not a procedural courtesy — it is the mechanism through which ownership is created.

Diagram 14.8: Team Ritual Design: Simple, Genuine, Owned

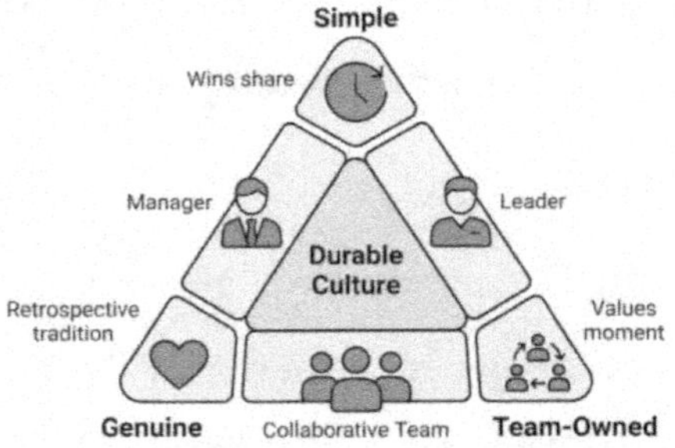

Integrating these three elements creates rituals that stick and strengthen team dynamics over time.

15.9 Manager's Checklist for Building Culture and Engagement

- Audit your four cultural levers monthly: what have you modeled, celebrated, tolerated, and refused in the past thirty days? Align them with the culture you intend to build.
- Facilitate a team values articulation session within your first ninety days with any new team. Revisit the values annually and whenever the team composition changes significantly.
- Assess psychological safety quarterly through direct conversation and pulse data. When safety scores decline, treat it as an urgent operational issue, not a developmental observation.

- Calibrate challenge levels for each team member in quarterly one-on-ones. Identify the specific skills or experiences that would constitute growth for each person in the next ninety days.

- Audit your recognition patterns quarterly to ensure equity in visibility across the team. Track who has been recognized and for what, and identify gaps.

- Establish at least two hybrid-equitable meeting practices — rotating locations, asynchronous input channels, or equivalent visibility mechanisms — before assuming remote team members are fully included.

- Monitor workload distribution using available project management and AI tools to make invisible overload visible before it becomes burnout.

- Design at least one team ritual collaboratively within the first sixty days of leading a team, and protect it from calendar pressure. Consistency is what makes rituals meaningful.

- Use AI pulse survey data as a conversation starter, not a performance indicator. Bring declining trends to the team rather than managing them privately.

15.10 **What Sustains Culture Over Time**

Culture is not built in a single initiative and then maintained on autopilot. It is sustained through the accumulated weight of daily decisions, consistent behaviors, honest conversations, and the willingness to hold the line on what matters even when it is inconvenient. The teams that maintain strong culture through personnel changes, leadership transitions, and organizational disruptions are the teams whose managers have been consistent enough that the culture has become genuinely collective — embedded in the team's identity rather than dependent on any single person. The path to that collective ownership is long and requires sustained, deliberate attention.

Your ultimate goal in culture-building is to create a team that carries its culture without you. Not because you want to disengage, but because a team whose culture depends entirely on the manager is fragile, and a team whose culture belongs to its members is not. Build the rituals, articulate the values, model the behaviors, and celebrate the moments — then watch as those elements take root in the team's collective identity and begin to sustain themselves. That is what a healthy culture looks like, and it is among the most lasting contributions a manager can make. Culture is the legacy that outlasts every project deliverable, every performance review, and every organizational restructuring. Build it deliberately. Tend it consistently. And watch what becomes possible when a team is genuinely, durably, authentically engaged.

16 The Future-Ready Manager — Continuous Growth and Innovation

16.1 Scenario

It is performance review season, and you are completing the self-assessment section. The question asks you to describe your professional development goals for the coming year. You type a sentence about improving your stakeholder communication, then delete it; type something about deepening your technical knowledge, then delete that too. The honest answer — the one you will not submit — is that you are not entirely sure what skills you need to develop because you are not entirely sure what a manager's job will look like in three years. The AI landscape is shifting fast enough that tools you adopted six months ago are already being superseded. Roles on your team that seemed stable are being redefined. The skills that made you successful as an individual contributor feel increasingly distant from the capabilities your manager is evaluating you on. The skills that made you effective in your first two years as a manager are no longer sufficient on their own.

This uncertainty is not a personal failing — it is the defining professional condition of this moment. But the managers who will thrive in the coming decade are not the ones who eliminate this uncertainty. They are the ones who develop a systematic relationship with it — who turn continuous learning into a genuine leadership practice, who build teams capable of innovating amid ambiguity, and who remain genuinely curious about what is coming rather than defensively committed to what has been. The future-ready manager is not the one with the most answers. It is the one with the best questions and the discipline to keep asking them. This chapter is a practical guide to building that disposition and the practices that support it.

Diagram 15.1: The Future-Ready Manager: Four Core Capacities

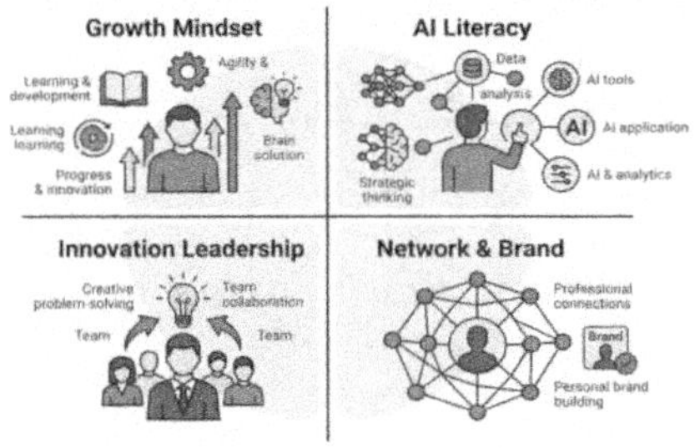

Cultivating these capacities empowers managers to thrive in an ever-evolving business landscape.

16.2 Why It Matters

The half-life of professional skills is shrinking. Capabilities that required years to develop are being automated; entirely new capabilities are being demanded of roles that did not previously require them. For managers, this creates a compounding challenge: you must continuously develop your own capabilities while simultaneously preparing your team members for skills and roles that the organization has not yet formally defined. The manager who addresses only one side of this challenge — developing personally while neglecting team preparation, or investing in the team while stagnating personally — will find that both sides eventually suffer from the imbalance.

The cost of falling behind is not theoretical. Managers whose development stagnates become organizational liabilities — not because they become bad at what they already know, but because what they already know becomes an insufficient foundation for the decisions they are being asked to make. The manager who last updated their industry's mental model three years ago is navigating today's decisions with yesterday's map. In a slowly changing environment, that gap is manageable. In an environment reshaped by AI at the current pace, it is not. The decisions that matter most — about which tools to adopt, which workflows to invest in, which capabilities to develop, how to position the team relative to emerging organizational needs — require current knowledge, not historical fluency.

There is also an opportunity dimension to this challenge. The managers who engage proactively with emerging AI tools, who position their teams as pilots for new capabilities, who build external visibility as practitioners navigating this transformation honestly — these managers are disproportionately likely to be recognized, promoted, and given the kinds of opportunities that accelerate careers. The future-ready manager is not just protected against obsolescence; she is actively advantaged by a willingness to engage with what is coming before everyone else does. Readiness is not a defensive posture — it is a competitive one.

16.3 The Growth Mindset for Managers

The concept of growth mindset, developed by Stanford psychologist Carol Dweck, describes the belief that fundamental qualities — intelligence, abilities, character — can be developed through dedication and effort. The fixed mindset, its counterpart, holds that these qualities are stable traits that you either have or you don't. While mindset research applies to all learners, its implications for managers are specifically powerful. Because managers operate in a role where authority and expertise are expected, the temptation to protect the appearance of competence at the expense of genuine learning is unusually strong. The manager's social position creates specific fixed-mindset pressures that do not exist in the same form for individual contributors.

A manager operating from a fixed mindset avoids situations where their lack of knowledge might be visible. They position themselves as the person with answers rather than the person asking questions. They interpret failure as a judgment on their capability rather than as data about a specific approach. They respond to team members who are more technically current with defensive distance rather than genuine curiosity. These behaviors are understandable — they reduce the discomfort of not knowing. Still, they are profoundly costly in an environment where the pace of change guarantees regular encounters with things you do not yet know. A fixed-mindset manager in a rapidly changing environment becomes an obstacle rather than an asset within months.

A manager operating from a growth mindset treats not-knowing as the starting point of learning rather than a condition to be hidden. They ask questions in front of their team and model intellectual humility as a leadership norm. They engage with failure as an organizational learning opportunity rather than a personal verdict. They actively seek out team members and peers whose knowledge exceeds their own and treat those interactions as development rather than as a competitive threat. These behaviors compound over time: the growth-mindset manager knows more and knows it better each quarter, while the fixed-mindset manager stands still and watches the knowledge gap widen.

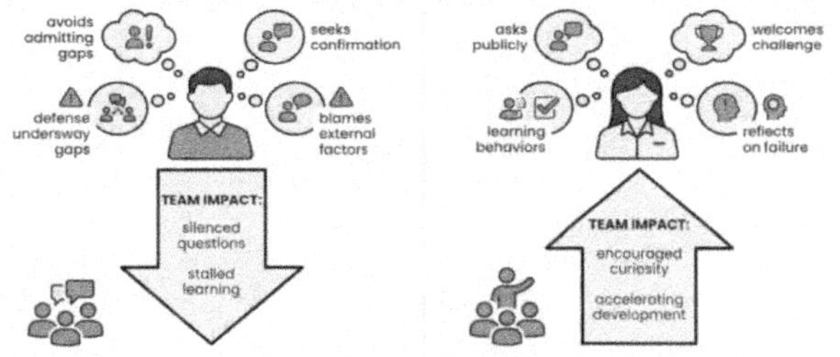

Embracing a growth mindset unlocks team potential, while a fixed mindset creates a culture of stagnation and silence.

16.3.1 Learning as a Leadership Practice

The distinction between managing and leading is nowhere more apparent than in professional development. Managers who treat their own learning as a private matter — something they do quietly when no one is watching — miss the cultural signal their learning behavior sends. Leaders who treat their learning as a visible, ongoing practice create permission for the same behavior throughout their team. The team takes its cues from the manager in this domain as in all others: if the manager is visibly learning, learning is visibly valued.

Making your learning visible does not mean sharing every article you read or narrating every insight you have. It means being transparent about what you are working to understand, referencing your learning in conversations with your team, and creating an explicit connection between your development and the team's evolving needs. "I've been spending time this month understanding how to evaluate AI vendors, because that decision is coming for us, and I want to make it well," is a leadership statement. It models learning as purposeful, connected to team needs, and something leaders do — not something they have already finished. It also signals that the decision process matters, not just the decision.

Learning as a leadership practice also means creating time for it in the organization. Managers who declare learning important but never allocate calendar time for it are modeling that learning happens in the gaps between real work, which is the definition of deprioritization. Building protected time for your own development, and for your team's development, signals that it is treated as the strategic investment it is. The calendar is the most honest signal of organizational priorities: if learning does not appear on it, it is not actually a priority, regardless of what the values documents say.

16.4 Building Your Professional Development Plan

A professional development plan is not a document you complete for an HR system. It is a working tool for managing the most important investment in your long-term effectiveness: your own capability. The best development plans are specific enough to guide action, flexible enough to adapt as conditions evolve, and connected enough to your current role that development and delivery reinforce each other. A development plan that is disconnected from your actual work is an aspirational document; a development plan that is embedded in your real challenges and real decisions is a learning system.

16.4.1 The Four Development Domains for Managers

A well-constructed development plan for a manager addresses four domains simultaneously: leadership capabilities (how you lead people), domain expertise (what you know about the industry and function you lead), AI and technology fluency (how you work with and evaluate the tools reshaping your environment), and external presence (how you are perceived and connected beyond your immediate organization). Each domain contributes distinctly to your effectiveness, and neglecting any one of them creates a vulnerability that eventually becomes visible.

Most development plans focus almost exclusively on the first two domains, neglecting the latter two. This was an acceptable gap ten years ago; it is not today. AI and technology fluency is now a leadership capability — managers who cannot evaluate an AI tool's claims, identify its risks, understand its data requirements, or make informed adoption decisions are operating with a material competency gap. And external presence — the network, the brand, the professional visibility that gives you access to information, opportunities, and perspectives beyond your organization — has always mattered for career development but has become dramatically more actionable through the platforms and communities that now exist for professional practitioners.

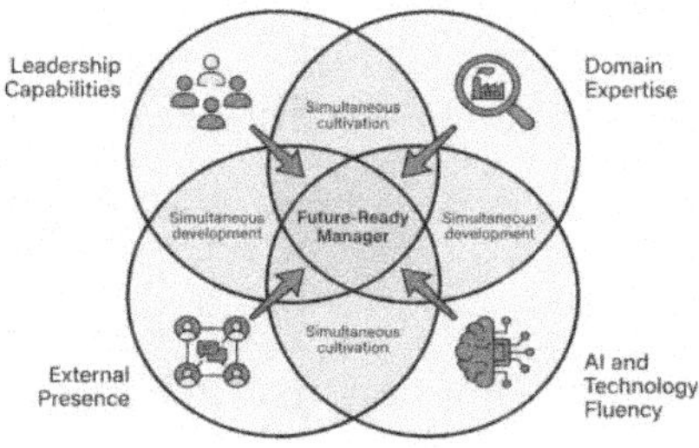

16.4.2 Setting Operational Development Goals, Not Aspirational

The failure mode of most development plans is abstraction. "Improve my communication skills" is not a development goal — it is a category. "Conduct three difficult feedback conversations using the SBI framework and debrief each one with a trusted peer" is a goal that can be executed, measured, and reflected upon. "Build my AI literacy" is a category. "Complete a hands-on evaluation of three AI project management tools against our team's current workflow and document the findings for a team discussion" is a goal that produces real learning as a byproduct of real execution. The specificity of the second form in each pair is not cosmetic — it is the characteristic that makes development actually happen.

The discipline of making development goals operational is the same discipline required to make team goals effective. Specificity creates accountability. Specificity also creates learning — when a goal is concrete enough to execute, the execution produces experience, and experience produces actual development in a way that abstract goals rarely do. A vague goal can remain perpetually in progress without ever requiring the uncomfortable encounter with reality that produces growth. A specific, dated, mechanism attached goal cannot be avoided in the same way.

Attach each development goal to a timeline and to a specific mechanism: a course, a project, a conversation, a reading commitment, a mentoring relationship, or a conference. The mechanism is not the goal — the outcome is. But without a mechanism, most development goals remain aspirational indefinitely. The mechanism is the bridge between intention and action, and the timeline is the accountability structure that keeps the bridge from remaining perpetually under construction. Review your development plan monthly, not annually, and revise it whenever your understanding of your most important gaps changes.

16.5 Staying Current with Emerging AI Trends and Tools

One of the most common development mistakes senior managers make is treating AI literacy as a one-time acquisition. They attend a workshop, read a book, adopt two or three tools, and check the box. But the AI landscape is evolving at a pace that makes point-in-time learning obsolete within months. Staying current requires a sustained information practice — a deliberate, regular process for monitoring what is changing, evaluating what is relevant, and updating your mental model accordingly. The manager who does this consistently will always be ahead of the manager who relies on episodic learning interventions.

16.5.1 Building an AI Intelligence System

An AI intelligence system is not a feed of every AI news article. It is a curated, maintained set of sources that covers the developments most relevant to your role and your industry, at a frequency and depth that is sustainable given your other obligations. Most effective AI intelligence systems for managers combine three layers: a small set of trusted general sources that cover the AI landscape broadly, a focused set of practitioner-level sources specific to your industry or function, and a human network of peers and specialists who surface things your algorithmic feeds miss.

For the general layer, practitioner-focused publications, research organization newsletters, and well-curated AI practitioner communities offer signal-to-noise ratios more useful than general technology news. For the industry layer, your professional associations, sector-specific research, and peer manager communities in your domain are often more valuable than the general AI press. For the human network layer, a small set of relationships with people who are slightly ahead of you in AI adoption — peers at other organizations, specialists within your own, thoughtful early adopters in your network — provides the early warning system that no publication can fully replace. Human networks surface early-stage, context-specific intelligence that algorithmic feeds cannot generate.

The goal of the intelligence system is not to know everything — it is to maintain enough situational awareness that you can recognize a relevant development when you see it, ask informed questions when evaluating new tools, and avoid being blindsided by changes that affect your team's work. The test of a good intelligence system is not whether you know every AI development, but whether you can identify the ones that matter to your team before they become mainstream knowledge. That margin of early awareness is the practical value of staying systematically up to date.

Diagram 15.4: The Manager's AI Intelligence System

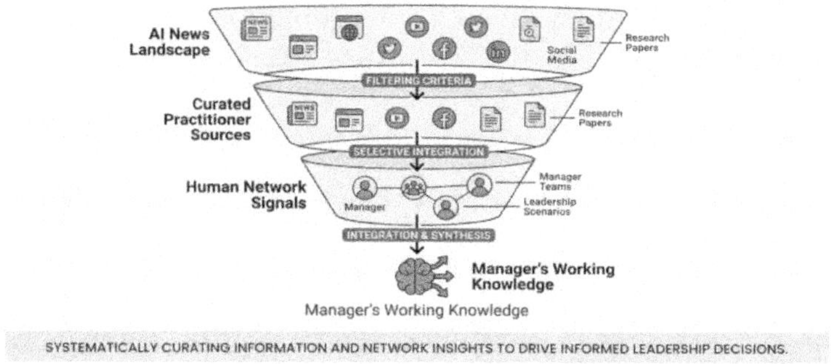

16.5.2 Evaluating New AI Tools Without Hype

The AI vendor landscape is saturated with claims. Every tool promises transformational productivity gains, every product roadmap is full of features that will arrive in the next quarter, and every demo is optimized to show the best possible scenario with ideal conditions that do not exist in your actual workflow. Learning to evaluate AI tools with appropriate skepticism — not cynicism, which closes off genuine opportunity, but skepticism, which demands evidence — is a critical management skill. The manager who evaluates every tool against concrete evidence of value in comparable contexts will make dramatically better adoption decisions than one who evaluates tools against their own excitement.

A useful evaluation framework for AI tools covers five dimensions. First, problem specificity: does this tool solve a problem that is genuinely significant in your current workflow, or does it solve a problem you don't actually have? Second, evidence quality: are the productivity claims based on controlled trials in comparable environments, or on cherry-picked testimonials and vendor-provided case studies? Third, integration reality: how does this tool actually fit into the current technology stack, and what are the real costs of integration and maintenance? Fourth, data and privacy considerations: what data does this tool require access to, how is that data used, and does its use comply with your organization's privacy and security requirements? Fifth, adoption likelihood: given your team's current skill levels, attitudes, and workload, is it realistic to expect adoption at a level that would generate the claimed benefits?

Running this framework does not require technical expertise. It requires clear thinking, honest assessment, and the discipline to ask the second and third questions rather than accepting the first answer. The first answer — the vendor's pitch — is always compelling. The second and third questions determine whether the tool will actually work in your environment. Developing the habit of asking them consistently is one of the most practical manifestations of AI literacy for a non-technical manager.

16.6 Fostering a Culture of Experimentation and Innovation

The teams that will perform best over the next decade are not necessarily the ones with the most sophisticated AI tools — they are the ones with the strongest culture of thoughtful experimentation. Experimentation cultures are those in which trying new approaches is encouraged, failure is treated as information rather than an indictment, learning is drawn from every attempt, and insights from pilots are systematically applied to production decisions. This culture does not emerge from declared values; it emerges from a consistent pattern of management behavior that makes experimentation safe, visible, and rewarded.

Building this culture is a management responsibility. Teams do not develop experimentation cultures spontaneously — they develop them because their manager created the conditions, absorbed the organizational risk of early failures, and modeled the intellectual disposition required. The manager who celebrates a pilot that produced

432

important negative results as loudly as one that produced positive results sends a message that matters: the organizational goal is learning, and learning requires the willingness to be wrong. That message, delivered consistently over time, produces teams that try things, learn from them, and compound that learning into a genuine competitive advantage.

16.6.1 The Innovation Protocol

A practical innovation protocol gives your team permission to experiment within a structure that manages risk, captures learning, and connects pilots to real decisions. The protocol has four elements: a proposal mechanism, a pilot boundary, a measurement commitment, and a learning review. These four elements together create the scaffolding for organized experimentation — systematic enough to be trustworthy, lightweight enough to be used regularly, and explicit enough to produce learning rather than just activity.

The proposal mechanism is a regular, lightweight process through which team members can propose experiments — new tools to try, new workflows to pilot, new approaches to existing problems. The mechanism should be low-friction enough that people actually use it, and it should have a predictable response: proposals are reviewed, given a yes-or-hold-with-rationale within two weeks, and if approved, assigned to a named owner. The predictability of the response is as important as the response itself: a proposal process that disappears into a black hole kills the proposal culture faster than rejection does.

The pilot boundary defines the scope and duration of the experiment: what will be tried, by whom, for how long, with what resource investment. Boundaries are critical because they contain risk and create a natural evaluation point. An experiment without a defined boundary tends to either die quietly or expand indefinitely — neither produces useful learning. The measurement commitment specifies, before the experiment begins, how success will be evaluated. This commitment prevents the common failure mode of retroactive rationalization, where experiments that do not produce the desired results are deemed successful by redefining the success criteria after the fact.

The learning review is a structured conversation at the end of the pilot that asks, regardless of outcome: what did we learn? What would we do differently? Should this be adopted, modified, discontinued, or studied further? The learning review is the most frequently skipped step in organizational experimentation — and the most important, because it converts experience into organizational knowledge. An experiment without a learning review is an expenditure; an experiment with one is an investment. The distinction is the difference between a team that tries things and a team that actually learns from what it tries.

Innovation is a disciplined cycle, not a spontaneous event.

16.6.2 Preparing Your Team for Roles That Don't Exist Yet

One of the most genuinely difficult management responsibilities of this moment is developing team members for futures that are not yet fully defined. The specific AI tools, platform capabilities, and workflow architectures that will define your industry in five years are not yet fully known. But certain underlying capabilities have historically been durable across technological transitions: the ability to frame complex problems, to communicate clearly across technical and non-technical audiences, to build and maintain productive working relationships, to evaluate evidence critically, and to learn new tools and approaches quickly. These meta-capabilities do not become obsolete when the tools change; they are what make adoption of new tools possible.

When planning development investments for your team members, anchor their growth in these durable capabilities while simultaneously exposing them to the emerging tools most likely to be relevant. This dual investment — deep capability plus current tool fluency — creates the combination most likely to remain valuable across multiple technological iterations. A team member with strong foundational capabilities and comfort with current AI tools can adapt to the next generation of tools more quickly than one with either dimension alone.

Create deliberate exposure to adjacent capabilities and roles. Assign team members to cross-functional projects to expose them to diverse perspectives and skill sets. Invite specialists — in AI, in your industry's emerging areas, in adjacent functions — to share what they are seeing. Facilitate internal learning sessions where team members teach each other the new skills they are developing. These practices expand the team's collective map of what is possible and build the cross-domain curiosity that is among the most durable professional assets any team member can possess in an era of continuous change.

16.7   Building Your Leadership Brand and Network

Leadership brand is the accumulated impression others form of you as a professional — your expertise, your values, your approach to problems, and your impact. Brand is not self-promotion; it is the reputation you build through the quality and consistency of your work, your communication, and your engagement with the professional community over time. In an era when professional networks are largely digital, and the work of managing AI-augmented teams is being navigated by practitioners everywhere simultaneously, an external leadership brand is both more accessible to build and more valuable than ever. The platforms and communities available to practitioners today would have required decades to build organically through in-person professional relationships twenty years ago.

16.7.1 Building Visibility Through Contribution

The most durable leadership brands are built through genuine contribution: sharing what you have learned, engaging honestly with the challenges you are navigating, and helping others who are facing similar situations. This is different from personal marketing. It is professional generosity that happens to build visibility as a byproduct. The distinction matters both ethically and practically: audiences for professional content are sophisticated enough to recognize the difference between self-promotion dressed as insight and genuine practitioner knowledge, and they respond to the latter with a depth of engagement that the former cannot produce.

The forms of contribution available to managers at every career stage are more varied than they may appear. Writing briefly about a specific challenge you navigated on a professional platform creates a record of your thinking and an invitation for conversation. Speaking at a professional event — even a small one, even virtually — demonstrates your ability to communicate and positions you as a practitioner. Participating thoughtfully in online communities around AI, management, or your industry adds to the collective knowledge and introduces you to people at organizations you might never otherwise encounter. Mentoring someone more junior shares your hard-won knowledge and builds your reputation as someone who invests in others. All of these activities produce the byproduct of visibility, but that byproduct is most valuable when it is genuine.

None of these activities requires claiming expertise you do not have. In fact, the most authentic and effective professional content is often the most honest: "Here is the specific challenge I am working through, here is what I have tried, and here is what I still do not know." Practitioners at every level of experience find that kind of honesty more valuable than the polished success narrative, because it reflects the experience of navigating complex, uncertain work. The professional community you most want to be part of values honest engagement over confident performance.

16.7.2 Network as Intelligence Infrastructure

Your professional network is not just a career tool — it is an intelligence infrastructure. The people in your network are your early warning system for trends that have not yet reached mainstream coverage and your first-call resource when you face a challenge with precedent in another organization. Your reality check when you are considering a decision that benefits from a perspective outside your immediate context. A well-maintained network provides information that no publication, course, or internal resource can replicate, because it is specific to your situation and delivered by people who know you.

Building a useful network requires intentionality about who is in it. A network composed entirely of people with similar backgrounds, in similar roles, at similar organizations provides comfort but not much in the way of intelligence diversity. A network that includes people across industries, at different career stages, with different functional specializations, and geographic perspectives provides the richness of insight that genuinely informs better decisions. Diversity in a professional network is not a social justice imperative — it is an epistemological one: different perspectives illuminate different dimensions of complex problems, and the manager who can access multiple perspectives makes better decisions.

Maintaining a network requires a lightweight but consistent investment — periodic reach-outs, sharing relevant information when you encounter it, congratulating genuine achievements, and responding promptly when others reach out to you. The manager who only activates their network when they need something from it will find it much less responsive than the manager who tends it continuously as a genuine professional community. Reciprocity is the operating norm of useful professional networks, and managers who give more than they take build the most responsive and valuable networks over time.

Diagram 15.7: The Manager's Professional Network Map

Diverse networks provide critical insights and opportunities for modern managers.

16.8   The AI-Forward Manager: Your 90-Day Action Plan

Growth intentions that do not translate into dated, specific actions rarely survive contact with a full calendar. The following 90-day action plan provides a structured scaffold for beginning the journey toward future readiness. It is designed to be realistic for a manager with a full operational workload — not requiring large blocks of dedicated development time, but requiring consistent small investments over twelve weeks. The compounding effect of consistent small investments in development outperforms sporadic large ones over any meaningful time horizon.

16.8.1 Days 1-30: Assess and Connect

In the first thirty days, the goal is assessment and connection. Conduct an honest audit of your current capabilities across the four development domains: leadership, domain expertise, AI fluency, and external presence. For each domain, identify your strongest current capability and the most significant current gap. Do not aim for comprehensiveness — aim for accuracy. The most useful development plan is built on an honest assessment of where you actually are, not where you would like to be or where you feel comfortable presenting yourself.

Simultaneously, begin building or refreshing your AI intelligence system. Identify three to five sources that you will monitor consistently, and establish a reading or listening practice that integrates them into your existing routine without requiring additional blocked time. Subscribe, bookmark, schedule — whatever makes consistency realistic for you specifically. The system only works if it is maintained, and it is only maintained if it fits your actual workflow rather than an idealized version of your calendar.

Reach out to three people in your existing network who are slightly ahead of you in AI adoption or whose perspective on the future of your industry you respect. Ask for a conversation — not to ask for anything, but to compare notes. These conversations are among the highest-value development investments available to a manager, and they are almost always reciprocally beneficial. The person you reach out to for perspective will often gain as much from the conversation as you do.

16.8.2 Days 31-60: Pilot and Share

In the second month, move from assessment to action. Choose one AI tool or capability that you assessed as relevant to your team's work and run a formal pilot using the innovation protocol: define the experiment, set the measurement commitment, run it for twenty-one to thirty days, and conduct a learning review. The specific tool matters less than the discipline of the process — the goal is to develop your own capacity to run systematic experiments, not to find the perfect tool on the first attempt.

Simultaneously, begin making your learning visible to your team. Reference what you are working to understand in one-on-ones and team meetings. Create one structured learning moment for the team — a thirty-minute session where you share what you have learned from the pilot, invite their perspective, and model the combination of informed conviction and genuine openness to revision that is the hallmark of a learning leader. The session demonstrates that learning is a collective activity that happens in the open, not a private one that happens before you present conclusions.

Start one external contribution, however small. Write one professional post about a specific challenge you have navigated. Reply thoughtfully to a discussion in a professional community. Share an insight with your network that you believe is genuinely useful. The specifics matter less than beginning the practice. The discomfort of the first contribution is real and is the same discomfort that every practitioner who eventually builds a valuable professional presence has navigated. Move through it rather than avoiding it.

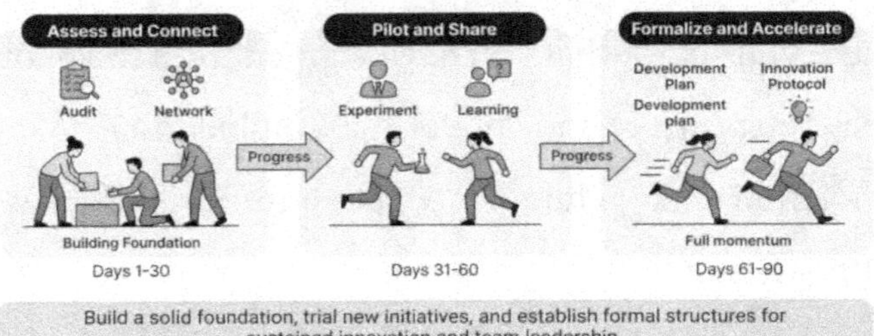

16.8.3 Days 61-90: Formalize and Accelerate

In the third month, formalize the practices that have proved useful and accelerate the ones that are building momentum. Document your development plan across the four domains with specific goals, mechanisms, and timelines — the version that is actually operational, not the version you submit to HR. Share it with a mentor or trusted peer who will hold you accountable to it. External accountability is not a sign of weak self-discipline — it is a recognition that development intentions, like all intentions, benefit from social structures that make follow-through more likely.

Establish the innovation protocol formally on your team—not as a grand announcement, but as a practical mechanism: here is how we will propose, pilot, measure, and review new ideas going forward. Run the first official experiment under the protocol, and use the learning review to model the behavior you want normalized. The formalization of the protocol is itself a signal: it communicates that experimentation is now a standard part of how the team operates, not an exceptional activity that requires special permission.

Identify one team member who is ready to take on a development stretch — a project, a learning commitment, or a visible opportunity — that would not have been available to them in the previous configuration of the team. Making that investment is not just good people development; it is evidence that the future readiness work you have done is translating into real opportunity for the people you lead. The team member who receives a stretch opportunity they did not expect becomes a concrete demonstration that the manager's development work has a downstream effect on the team.

16.9 Manager's Checklist for Continuous Growth and Innovation

- Conduct a quarterly audit of your four development domains: leadership capabilities, domain expertise, AI fluency, and external presence. Identify the most significant gap in each and design one specific action to address it.
- Maintain an AI intelligence system with at least two curated sources you review weekly. Log one relevant

trend or development per month and share it with your team.

- Run at least two formal innovation protocol experiments per quarter. Conduct a learning review for each outcome, regardless, and share the findings with the broader team.
- Make your learning visible in at least one team interaction per week — referencing what you are working to understand, what you tried recently, or what a pilot revealed.
- Build protected development time into your calendar — at minimum, two hours per week — and treat it with the same commitment you apply to a critical stakeholder meeting.
- Conduct at least two external network conversations per month with people outside your immediate organization. Prioritize diversity of perspective over familiarity.
- Contribute to at least one external professional community per quarter — a post, a conversation, a presentation, a mentoring relationship.
- Develop one team member per quarter for a stretch opportunity that expands their capability beyond their current role description.
- Review your 90-day action plan at day 30, day 60, and day 90. Revise it based on what you have learned, not what you planned to learn.

16.10 **Your Defining Advantage**

The manager's role has never been static, and it will not become static. Every decade brings new tools, new organizational models, new workforce expectations, and new demands on the people responsible for leading others through complexity. The managers who thrive are not the ones who found the perfect model and executed it faithfully — they are the ones who treated their own development with the same rigor and intentionality they applied to their teams', and who built the adaptive capacity to remain effective as the landscape continued to shift beneath them. Development is not something you complete; it is something you practice continuously, with increasing skill and increasing self-awareness about what you most need to work on.

The AI era does not diminish the value of the manager. It relocates it. The work that machines cannot do — the judgment calls that require contextual understanding, the human conversations that require genuine empathy, the ethical decisions that require values-grounded reasoning, the culture-building that requires sustained authentic presence — is precisely the work that becomes more important as automation handles more of the rest. The future-ready manager is not the one who competes with AI. She is the one who understands what AI enables, leads the humans navigating that enabling, and models the kind of continuous learning and genuine engagement with uncertainty that defines leadership in any era.

You have chosen a profession that requires constant renewal. That is not a burden — it is an invitation to keep growing, to stay curious, and to lead in a way that leaves every team you manage more capable, more confident, and more prepared for what comes next than they were before you arrived. The leaders who look back on their careers with genuine satisfaction are not those who did the same thing exceptionally well for forty years. They are the ones who kept learning, kept adapting, and kept bringing that growth to the people they were responsible for. That is the work. And it is worth doing well.

17 Your Leadership Journey Starts Now

You have covered a great deal of ground in this book. Fifteen chapters, five parts, and a single sustained argument: that the most effective managers in the AI era are not the ones who mastered the most technology, but the ones who developed the greatest human skills and then used technology to extend them.

It is worth pausing for a moment to name what you have built.

You started by confronting the identity shift—the moment when the skills that made you a high-performing individual contributor no longer suffice, and the entirely different work of leading people begins. You built a leadership foundation grounded in your values, your decision-making principles, and your understanding of what it means to be accountable for a team rather than for your own output.

You developed the core management competencies: how to communicate with the clarity and candor that teams need to function, how to delegate with enough precision that work comes back right, how to set goals that focus effort and drive performance rather than measure it after the fact. You learned how to hire with rigor, onboard with intention, and coach in ways that actually develop people rather than evaluate them.

You tackled the hardest parts of the role—conflict, difficult conversations, and the moments when the professional and the personal blur in ways that require both courage and care. You built a personal productivity system that protects your time and energy so you can show up fully for the people who depend on you.

Then you turned to AI—not as a shortcut, but as a precision instrument. You built the literacy to evaluate AI tools critically. You learned how to apply them to the specific management challenges you face: communication, delegation, feedback, data analysis, and decision support.

You explored what it means to make decisions with data without surrendering your judgment to algorithms.

You examined the larger forces shaping your context—digital transformation, organizational change, and the psychology of teams navigating uncertainty. You built a framework for cultivating the culture and engagement that make teams resilient rather than merely functional. And you closed with a vision of continuous growth: the manager who is still learning, still adapting, and still investing in their own development five years from now.

That is the complete picture of what you have built. The question now is what you do with it.

17.1   Your 90-Day Action Plan

The research on skill development is consistent: the half-life of learning without application is short. The managers who get the most from this book are the ones who convert its frameworks into specific actions in the weeks immediately following.

In your first thirty days, choose one competency from each part of the book and implement a single practice. Hold one coaching conversation using the structure from Chapter 7. Write one delegation brief using the framework from Chapter 4. Set up one AI-assisted workflow from Chapter 11. These are not grand transformations—they are small experiments that build evidence about what works in your specific context.

In days thirty-one through sixty, review what you implemented. What worked? What needed adjustment?

What did your team respond to, and what landed differently than you expected? Use this feedback loop to refine your approach. The goal is not to have all fifteen chapters running simultaneously—it is to build sustainable habits that compound over time.

In days sixty-one through ninety, identify the one area where you have the highest leverage for your team right now. It might be a communication issue, because a clarity problem is slowing execution. It might be coaching, because you have two direct reports with significant growth potential who are not being challenged. It might be AI adoption, because your team is behind the curve on tools that could materially improve their output. Whatever it is, go deep. Make that one area genuinely excellent before expanding your focus.

After ninety days, repeat the cycle.

17.2   On Continuous Learning

The AI era has a characteristic that makes continuous learning non-optional: the tools change faster than any single book can keep track of. What this book has given you is not a catalog of current AI tools—those will be different in eighteen months. What it has given you is a framework for evaluating tools, a set of questions to ask when a new capability appears, and a management foundation strong enough to make any tool useful.

Stay curious about what is emerging. Talk to peers who are experimenting with tools you have not tried. Read widely—not just about AI, but about organizational behavior, psychology, and the industries your team operates in. The managers who thrive in the long run are the ones who

treat their own development as a professional obligation, not
an occasional indulgence.

Find or build a community of practice. The managers
who grow fastest are rarely growing alone. They have peers
they talk to honestly about what is hard, mentors who have
navigated the terrain before them, and direct reports whose
candid feedback they actively solicit. None of that happens
by accident. It requires the same intentionality you bring to
any other management practice.

17.3   The AI Era Is Your Advantage

There is a version of the story about AI and management
that is fundamentally anxious: AI is coming, the role is
changing, uncertainty is everywhere, and no one is sure what
management even means in five years. That story is not
wrong about the facts—the uncertainty is real. But it draws
the wrong conclusion.

The AI era is an advantage for managers who have done
the work described in this book. When AI handles the
mechanical parts of management—the scheduling, the first
drafts, the data aggregation, the pattern recognition in
performance metrics—what remains is the work that only
humans can do: building trust, developing judgment,
navigating ambiguity, inspiring people to do their best work
in conditions that are not always ideal. That work becomes
more visible, more valued, and more differentiating as AI
takes on everything else.

You are entering an era where your human capabilities
as a leader are not being deprecated—they are being
amplified. Managers who understand this will build

extraordinary teams. The managers who do not will spend the next decade trying to keep up with tools instead of leading people.

You have done the work to understand it. Now lead.

www.ingramcontent.com/pod-product-compliance
Lightning Source LLC
Chambersburg PA
CBHW051409050726
47595CB00010B/3999